Mathematics in Industry

Volume 35

Mathematics in Industry focuses on the research and educational aspects of mathematics used in industry and other business enterprises. Books for *Mathematics in Industry* are in the following categories: research monographs, problem-oriented multi-author collections, textbooks with a problem-oriented approach, conference proceedings. Relevance to the actual practical use of mathematics in industry is the distinguishing feature of the books in the *Mathematics in Industry* series.

More information about this series at http://www.springer.com/series/4650

Hans Georg Bock • Karl-Heinz Küfer •
Peter Maass • Anja Milde • Volker Schulz
Editors

German Success Stories in Industrial Mathematics

Editors
Hans Georg Bock
Interdisciplinary Center for Scientific
Computing IWR
Heidelberg University
Heidelberg, Germany

Peter Maass
Center for Industrial Mathematics ZeTeM
University of Bremen
Bremen, Germany

Volker Schulz
Department of Mathematics
Trier University
Trier, Germany

Karl-Heinz Küfer
Fraunhofer Institute for Industrial Mathematics
Kaiserslautern, Germany

Anja Milde
Young Researchers and Diversity Management
Friedrich Schiller University Jena
Jena, Germany

ISSN 1612-3956   ISSN 2198-3283   (electronic)
Mathematics in Industry
ISBN 978-3-030-81457-1   ISBN 978-3-030-81455-7   (eBook)
https://doi.org/10.1007/978-3-030-81455-7

TABLE OF CONTENTS

PREFACE

Over twenty years ago, the slogan "Mathematics: Key technology of the future" became the vision and driving force for applied mathematicians working on real life industrial applications. Since then, mathematics has proved to deliver groundbreaking contributions for a wide range of applications; however, this is still only the tip of the iceberg. The huge technological potential of mathematics for problem-solving and algorithmic development is still just beginning to show its ability to produce industrial innovation. This book is a collection of success stories of industrial mathematics with the goal of giving inspiration to industrial and academic research units as well as decision-makers in industry and politics.

The contributions in this book demonstrate the power of mathematics as a technology. We show that mathematics can decisively contribute to the solutions of present and future economic, environmental, and societal changes. In selecting which stories to feature, we used the following criteria:

+ The industrial problems needed to be both relevant for the core business of the industrial partners and also ones which couldn't be solved using standard methods.

+ There had to be sufficient documentation of the domain-specific expertise of the industrial partners, including existing models, experimental data and evaluation criteria.

+ The use of novel mathematics was essential in solving the problem – for analyzing models, designing and optimizing algorithms, or efficiently implementing solutions beyond the present state of the art.

The selected contributions span different industrial sectors and methodologies but all involve two unique, underlying factors of mathematics. First of all, mathematics is the ultimate scientific discipline of rigor and precision. The concept of a mathematical proof is the basis for a reliable analysis of algorithms, for the correctness of physical-engineering models or for the statistical evaluation in scientific computing and data analysis. As a result, algorithms based on novel mathematical theory have shown a development similar to that of Moore's Law. And in all cases, such math-based algorithmic improvements outperform machine improvement, sometimes by significant factors.

Second, mathematics is a discipline of abstraction with the power to reduce any physical or engineering model to its essential core. This allows for an analysis of a model's properties in theoretical terms and a translation of the results back into their physical-engineering context. The analysis of a single abstract model typically covers the essential features of a variety of applications; hence, the path back from one new mathematical insight to applications is useful for additional physical/engineering models. As a consequence, mathematics is highly specialized in its own field but highly diverse and transdisciplinary in its applications.

In particular, those two factors play a decisive role when analyzing novel AI (artificial intelligence) concepts or general machine learning algorithms for data analysis. These approaches can be seen as implicit model adaptations allowing for modelling, simulation and optimization of highly non-linear, high-dimensional problems. Mathematics is the basis for merging the best of both worlds, i.e. for integrating model-based and domain-specific expert knowledge into such data-driven concepts. Moreover, it allows for determining the potential as well as the limitations of AI algorithms and it leads to novel network architectures. Mathematics is indispensable for obtaining optimal simulation results.

Before proceeding with the main content of this book, we want to acknowledge and emphasize that most of the following contributions have benefitted at least partially from the unique funding program 'Mathematics for Innovations' of the German Federal Ministry of Education and Research (BMBF). This funding program has existed for more than 25 years, and has created powerful collaborations between industry and mathematical research. It has allowed us to train several generations of industrial mathematicians and most importantly, it has led to substantial and sustainable industrial innovations.

MATHEMATICS MAKES THE IMPOSSIBLE POSSIBLE

Digitalization is the megatrend of the 21st century. Due to the exponential growth of computing power (Moore's law) and the miniaturization of chips, digitalization is now penetrating all areas of life and all industries. At the core of digitalization are algorithms and especially mathematical algorithms. In the last decades, their performance has also increased exponentially, in many cases even outpacing the growth in computing power. Without mathematics this incredible increase in performance would not have been possible.

Many industrial products, systems, and infrastructures – from electric motors to computer tomographs or traffic networks – rely on corresponding algorithms. Life without algorithms is hard to imagine. Today, they enable us to predict and test the properties and performance of products using virtual models long before they are built. For example, this allows early optimization of products and systems to realize continuously increasingly sophisticated, efficient and ecologically compatible products. Without appropriate simulation methods, the speed of today's innovation cycles could not be realized. Many technologies would not have been mastered. At the same time, corresponding virtual models can also be used during operation, e.g. to realize highly efficient control systems. Mathematics enables highly complex models to be run parallel to operation by means of sophisticated pre-calculations. Combined with continuous streams of sensor data and machine learning allows to optimize operation and service to a level not seen before.

The continuous prediction and optimization of characteristics, behavior, and performance, from early design to end-of-life, using digital information, data, and models is reflected in the vision of the digital twin. The digital twin and corresponding software solutions are also central technologies at Siemens AG. For example, the corresponding research and predevelopment is well funded, with investments in the high doubledigit millions. Examples include mathematical innovations[1] enabling the realization of industrial milling robots[2] or increasing the availability of motors [3].

Without mathematics, many products would be unthinkable, as the examples in this book impressively show. Mathematics is the key technology for digital twins and probably even the key technology for the 21st century itself.

Dirk Hartmann
Senior Principal Scientist Simulation &
Digital Twin, Siemens AG
Siemens Top Innovator
1st Deputy Chairman KoMSO

---

[1] https://new.siemens.com/global/en/company/stories/research-technologies/digitaltwin/passion-for-digital- twins.html

[2] https://youtu.be/2iIN-9Kno3o

[3] https://youtu.be/86vkjykbHRM

KEYNOTES

We live in the age of mathematics. Its influence pervades economic and social activity and its influence and impact are intense. Our economy increasingly relies on innovation to provide a significant proportion of the productivity gains required to support rising standards of living. Mathematics is playing an ever-expanding role in generating innovation and impact and, via knowledge exchange, is adding substantial social and economic value to our economy. This has been quantified in a number of reports published in several European countries, demonstrating that the economic contribution of mathematics represents around 25 – 30 % of a country's national income. In addition to the evidence from the reports on the economic impact, recent data may be used to show that mathematical research produces an outstanding rate of return on investment as compared to other disciplines like engineering, chemistry and physics[1].

Every day the mathematical sciences are used to solve otherwise intractable problems. We rely on cryptography securing our transactions over the internet and the optimal allocation of scarce resources, such as the radio spectrum which allows our mobile phones to work in crowded areas. Mathematics underpins numerous scientific, technical and social advances that improve health and raise living standards. Genetic analysis relies on statistical methodologies, allowing improvements in human, animal and plant health. Machine learning, artificial intelligence (AI) and data science are dependent on mathematics to find patterns in complex datasets and to explain the behaviour of deep neural networks. The risk of pension and investment funds is managed and minimised using mathematical models and actuarial science. Many more examples can be given here.

This booklet provides a number of impressive success stories setting out a very powerful and well-argued case for extra investment and new institutional initiatives that can accompany it. Taken together, they could be fundamental elements of the drive for productivity growth, innovation and a redefinition of the role of mathematics in a world which is seeing very rapid changes in international and technological structures. New mathematical understanding does not come out of the ether. It requires investment in the pure mathematics that underlies all the rest, in the applications working with partners and other disciplines, in the people, particularly the young who will take it forward, and in understanding of mathematics from the top CEOs and ministers to those in the more technical areas who will do the 'hard graft.'

The success stories in this booklet make a very powerful case for investing in mathematics. This is the age of mathematics and its influence will become still more intense. It is a discipline in which Europe can shine and lead. Now is the time to invest in its future.

Wil Schilders
President of EU-MATHS-IN (European Service Network of Mathematics for Industry and Innovation)
Officer at large, ICIAM (The International Council for Industrial and Applied Mathematics)
Director of Dutch National Platform for Mathematics
Professor of Scientific Computing for Industry, Eindhoven University of Technology

---

[1]"The era of mathematics", page 6

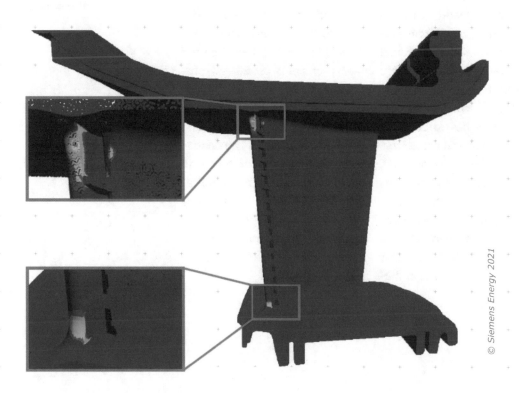

© Siemens Energy 2021

Hazard density for a turbine vane calculated with a poisson point process

© Springer Nature Switzerland AG 2021
H. G. Bock et al. (eds.), *German Success Stories in Industrial Mathematics*,
Mathematics in Industry 35, https://doi.org/10.1007/978-3-030-81455-7_1

FROM PROBABILISTIC PREDICTION OF FATIGUE LIFE TO A NEW DESIGN APPROACH FOR GAS TURBINES

Gas turbines are used in aviation and energy production. As efficiency of gas turbines increases with firing temperatures, the hot gas components of a gas turbine are subject to extreme termo-mechanical load. Activation and deactivation cycles from start to landing of an aircraft or during the temporal activation of a gas power plant to stabilize an energy grid drive mechanical fatigue, which therefore has to be taken into account in design procedures and operation and service plans. Fatigue life however is not predictable in a deterministic fashion. Even under lab conditions, material test specimens show a scatter of fatigue life by a factor 10 of the average life. The probabilistic assessment of risk during operation therefore is of highest importance. In almost 10 years of continuous collaboration, new design procedures for the probabilistic calculation of risks for low cycle fatigue (LCF) have been developed by Siemens. The mathematical modeling, based on the theory of point processes, provided a sound theoretical foundation that was complemented with intense research and testing in materials science. The continuous confrontation of modeling approaches with the requirements of industrial applicability and experimental data triggered innovation in materials science, mathematics and gas turbine design and operation.

TILMANN BECK
Technical University of Kaiserslautern

HANNO GOTTSCHALK
University of Wuppertal

ROLF KRAUSE
Universita della Svizzera Italiana

PARTNERS

GEORG ROLLMANN, SEBASTIAN SCHMITZ **and** LUCAS MÄDE, Siemens

Industrial challenge and motivation

Gas turbines are used for jet propulsion and power production. In aviation, gas turbine technology provides compact, lightweight and efficient engines. In power production, the specific CO_2 output per produced energy unit is $1/3$ rd of that of lignite power plants, if the gas turbine is run in a combined cycle with a steam turbine that uses the gas turbines' exhaust gas. Due to their quick start capabilities, gas fired power plants also play a role in stabilizing the energy grid in times of high influx of volatile renewable energy. At the same time, the demand for top efficiency remains an important driver for gas turbine development. Apart from GE, Mitsubishi Heavy Industries and Siemens offer technology for power plants operating at over 60% efficiency. As such efficiencies can only be achieved with firing temperatures way above the melting point of the engineering materials, sophisticated cooling technology is needed. Nevertheless, extreme thermal and mechanical loads during activation, deactivation and operation limit the life of the components which has to be thoroughly considered in design and and maintanancs. Quick start operation regimes lead to low cycle fatigue, in particular. The time to the initiation of a fatigue crack however is highly stochastic and the component safe life time rather is a random number than a deterministic value. Even under lab conditions, the maximum scatter band in lifetime is up to a factor 10 larger than the median life under the same nominal test conditions. Therefore, probabilistic design procedures and a proper calculation of probabilities of crack initiation is not only much more adequate given the experimental findings, but also provides more qualified information for risk assessment in design and maintenance The challenge in setting up a probabilistic design approach towards fatigue life has been taken up by Siemens (L. Mäde, S. Schmitz and G. Rollmann) in a joint effort with materials scientists from TU Kaiserslautern (T. Beck and B. Engel) and Jülich Research Center (T. Seibel) and Mathematicians from the University Lugano (R. Krause) and University of Wuppertal (H. Gottschalk, N. Moch and M. Saadi) in a series of research projects under the auspices of the

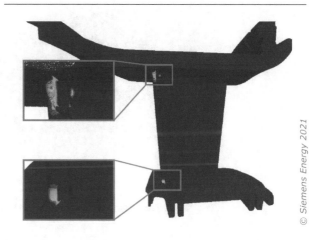

© Siemens Energy 2021

Figure 1: Hazard density for a turbine vane calculated with a poisson point process involving notch support factors

'AG Turbo' research collaboration co-funded by the Federal Ministry of Economics and Energy (BMWi) and industry partners within the AG Turbo. The scope of the first research period between 2011 – 2013 was an effective probabilistic model that is capable to provide probabilistic calculations of the risk of crack initiation over time for turbo-machinery components with complex geometry. In a second period 2012 – 2015, some consequences of probabilistic life prediction were exploited, namely the fact that shape sensitivities can be calculated for probabilistic – but not deterministic – component life. Application areas lie in design and the setting of geometric tolerances in production. This research and development (RnD) direction is further elaborated by the ongoing BMBF-funded GIVEN project (www.given-project.de).

In the third period with projects from 2014 – 2017 and 2019 – 2022 a better understanding of the impact of materials' heterogeneous micro structures as a root cause of the scatter of LCF life time is investigated and included in multi-scale approaches to probabilistic fatigue life calculation. This also gives interesting insight in the impact of parameters in production, as e.g. cooling and solidification rates in the casting process influence the micro-structure, to the safe life of a heavily loaded turbine component.

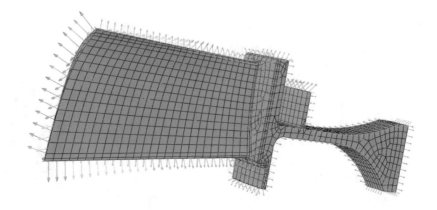

Figure 2: Shape sensitivities for failure probability calculated by the adjoint method

Mathematical and interdisciplinary research

The probabilistic models that have been proposed in the joint research effort are based on the theory of spatio-temporal point processes[14][7]. Indentifying a crack initiation event with a point on the component's surface x and a time t, we parameterize all configurations of crack initiations as (x, t) – pairs. A random crack initiation history then is given by a random configuration η that consists of collections of such pairs. Defining $\eta(A \times I)$ as the number of crack initiation events in a surface region A that happened in a time interval I, we obtain as the probability of survival without cracks

$$S(t) = P\left(\eta\left(\partial\Omega \times [0, t]\right) = 0\right)$$

$$= \exp\left(-\int_{\partial\Omega}\left(\frac{t}{N(\sigma)}\right)^m dA\right)$$

where is the entire surface of the component. In the second equation we already used a special Poisson point process that is based on the deterministic cycles to crack initiation as a scale parameter and a Weibull probability distribution with shape parameter $m > 1$. Here is the stress field that is calculated in practice as the numerical solution of the elasticity equation using finite elements. This basic model has been calibrated with data, tested experimentally and has been implemented a finite element post processor for the numerical

evaluation of survival or failure probabilities[12, 13], see Figure 1. Further extensions concern thermo-mechanical loads [13]and the modeling of notch support factors[10][9]. Interestingly, taking the probability of survival as an objective in shape optimization fosters new theoretical developments on the mathematical side: The probability of failure is a more singular objective functional as e.g. the compliance functional or tracking type objectives that are often used in shape optimization. This insight has lead to the development of theoretical foundations for shape optimization in connection with elliptic regularity theory[2, 7]. Mathematical results cover the existence of optimal shapes and detailed studies of the regularity of shape derivatives[1]. A crucial observation with survival probabilities is that they are essentially given as a surface integral, as opposed to a deterministic life given by a minimum life over all points on the component's surface. It therefore is possible to compute shape sensitivities using the adjoint method, see Figure 2. This program has been pursued in [6, 8] for complex 3d geometries. It is also shown that gradient based methods of mono and multi-criteria optimization can be effectively based on this approach, opening up new prospectives in multi criteria shape optimization [3].

The ongoing research on probablistic life calculation for low cycle fatigue does not satisfy itself with an empirical approach to the probability of failure like using the Weibull distribution. Instead, the

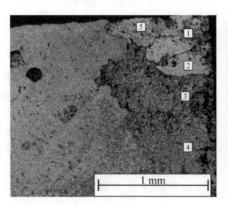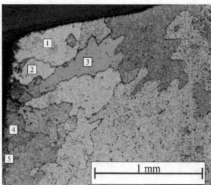

Figure 3: EBSD analysis of grains along a crack path

scatter in lifetime is derived from the random microstructure in the nickel based superalloys using multi scale modeling. These models are based on detailed experimental investigations using electon backscatter diffaction to measure the orientation of crystallographic planes in the neighborhood of locations of crack initiation, see figure 3. Several models have been compared and benchmarked, obtaining theoretical predictions of scatter bands of the size that is actually observed[4, 5, 11].

Implementation

The way forward for the RnD on the probabilistic modeling of low cycle fatigue has been characterized by the close integration of mathematical modeling and software prototyping, experimental work and the careful evaluation and review processes in industry, before tool development and new design procedures in industry could be based on the foundations laid in research. Test campaigns had to prove the validity of the mathematical models, going beyond the mathematical verification of their interior logical consistency. Based on this rigorous verification and validation, the probabilistic fatigue life methodology that has been developed in the first research phase has been implemented within Siemens to support decision-making in design and service of power plant components. In order to assess corresponding shape sensitivities in an efficient manner, the adjoint approach from the second collaboration period has

been incorporated in an in-house finite element tool suite as well. The current third phase of joint R&D focuses, on the one hand, on adjoint methods in multi criteria shape optimization, and on the other hand on refined multiscale models for the scatter in fatigue life.

Industrial relevance and summary

Within Siemens, the joint R&D effort has significantly contributed to the field of probabilistic life assessment. The resulting, more realistic service life assessments have allowed for additional customer value through increased sustainability by better utilization of power plant components. In addition, the use of probabilistic target functionals in design optimizations have led to improved component geometry definitions resulting in higher durability. Also because of this success-story, probabilistic design was presented as one of four key development areas during the Zurich forum of the Global Power and Propulsion Society in 2018.

Patents

» Procedure for the integrated prediction of failure probabilities for mechanical components due to material scatter and and manufacturing tolerances (with S. Schmitz and M. Saadi) submitted to EP office, file 102016221928.6 09.11.16.

Acknowledgements

The authors acknowledge Funding by the Federal Ministry for Education and Research via the GIVEN project under the grant no. 05M18PXA. This work has also profited form direct funding by Siemens and the Federal Ministry of Economic Affairs and Energy via grants 03ET2013I and 03ET7041J.

References

[1] L. Bittner. On shape calculus with elliptic pde constraints in classical function spaces, 2020.

[2] L. Bittner and H. Gottschalk. Optimal reliability for components under thermomechanical cyclic loading. *Control and Cybernetics*, 45, 2016.

[3] M. Bolten, H. Gottschalk, C. Hahn, and M. Saadi. Numerical shape optimization to decrease failure probability of ceramic structures. *Computing and Visualization in Science*, 21(1-6):1–10, July 2019.

[4] Engel, Mäde, Lion, Moch, Gottschalk, and Beck. Probabilistic modeling of slip system-based shear stresses and fatigue behavior of coarse-grained ni-base superalloy considering local grain anisotropy and grain orientation. *Metals*, 9(8):813, July 2019.

[5] B. Engel, T. Beck, N. Moch, H. Gottschalk, and S. Schmitz. Effect of local anisotropy on fatigue crack initiation in a coarse grained nickel-base superalloy. *MATEC Web of Conferences*, 165:04004, 2018.

[6] H. Gottschalk and M. Saadi. Shape gradients for the failure probability of a mechanic component under cyclic loading: a discrete adjoint approach. *Computational Mechanics*, 64(4):895–915, 2019.

[7] H. Gottschalk and S. Schmitz. Optimal reliability in design for fatigue life. *SIAM Journal on Control and Optimization*, 52(5):2727–2752, 2014.

[8] H. Gottschalk, S. Schmitz, T. Seibel, G. Rollmann, R. Krause, and T. Beck. Probabilistic schmid factors and scatter of low cycle fatigue (LCF) life. *Materialwissenschaft und Werkstofftechnik*, 46(2):156–164, Feb. 2015.

[9] L. Mäde, H. Gottschalk, S. Schmitz, T. Beck, and G. Rollmann. Probabilistic lcf risk evaluation of a turbine vane by combined size effect and notch support modeling. In *Turbo Expo: Power for Land, Sea, and Air*, volume 50923, page V07AT32A004. American Society of Mechanical Engineers, 2017.

[10] L. Mäde, S. Schmitz, H. Gottschalk, and T. Beck. Combined notch and size effect modeling in a local probabilistic approach for LCF. *Computational Materials Science*, 142:377–388, Feb. 2018.

[11] N. Moch. From microscopic models of damage accumulation to the probability of failure of gas turbines. 2019.

[12] S. Schmitz. *A Local, probabilistic Model for LCF*. Hartung-Gorre Verlag, 2014.

[13] S. Schmitz, H. Gottschalk, G. Rollmann, and R. Krause. Risk estimation for lcf crack initiation. In *Turbo Expo: Power for Land, Sea, and Air*, volume 55263, page V07AT27A007. American Society of Mechanical Engineers, 2013.

[14] S. Schmitz, T. Seibel, T. Beck, G. Rollmann, R. Krause, and H. Gottschalk. A probabilistic model for lcf. *Computational Materials Science*, 79:584–590, 2013.

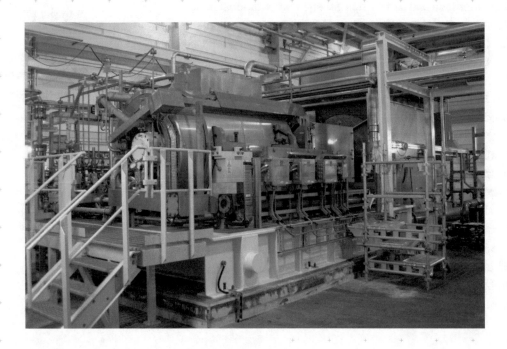

Optimizing torque control for a multi-megawatt compressor.

H. G. Bock et al. (eds.), *German Success Stories in Industrial Mathematics*,
Mathematics in Industry 35, https://doi.org/10.1007/978-3-030-81455-7_2

INCREASING THE RELIABILITY OF MULTI-MEGAWATT GAS COMPRESSORS

A homotopy framework for optimized real-time feedback

According to our industrial partner "interruption in the gas supply can mean financial losses that range into the hundreds of millions USD per year for large facilities". Fast methods and software for Nonlinear Model Predictive Control (NMPC) on embedded systems greatly improve the reliability of multi-megawatt gas compressors in comparison with less advanced control strategies in the face of power disturbances. The mathematical foundation for this success is a shift in numerical NMPC methods from a black box simulation and optimization paradigm to a structure exploiting homotopy framework, which facilitates the computation of optimizing feedback in the millisecond range even on computationally limited embedded hardware. The companies ABB, Gassco, and Statoil (since May 2018 called Equinor) cooperated in this project, using the mathematical and computational methods and software initiated at the Interdisciplinary Center for Scientific Computing at Heidelberg University and the KU Leuven Center of Excellence "Optimization in Engineering".

---

[1] Thomas Besselmann is currently with FHNW University of Applied Sciences and Arts Northwestern Switzerland, Windisch, Switzerland
[2] Sture Van de moortel is currently with ABB Excitation, Turgi, Switzerland

HANS GEORG BOCK
Interdisciplinary Center for Scientific Computing, Heidelberg University

CHRISTIAN KIRCHES
TU Braunschweig

ANDREAS POTSCHKA
TU Clausthal

THOMAS BESSELMANN
ABB Corporate Research, Baden-Dättwil, Switzerland[1]

STURE VAN DE MOORTEL
ABB Medium-Voltage Drives, Turgi, Switzerland[2]

Industrial challenge and motivation

Lightning strikes, winter storms, or ice buildup on power lines can lead to power disturbances, which severely affect the operation of 41.2 MW gas compressors at a large Equinor site in Kollsnes, Norway. Disturbances may lead to complete protection shutdown of the compressors, which halts the supply of gas. Restarting compressors and upstream facilities lasts from a few hours to days, resulting in tremendous financial losses. In order to keep the gas flowing, novel mathematical and computational methods are required to improve the control of the compressors to prevent them from stalling or going into surge.

Figure 1: When electrically driven multi-megawatt gas compressors go into surge due to disturbances in the power grid, large financial losses can be incurred by the sudden shutdown and slow restart of upstream facilities. Fast numerical methods for optimizing torque control based on a homotopy approach facilitate feedback rates in the millisecond range, which is required to perform partial torque ride-through for short power disturbances.

Problem description

Based on a mathematical model of the compressor dynamics, optimal control inputs to the gas compressors can be computed in reaction to electric signal measurements. It is important that the controls and resulting compressor currents adhere to safety constraints to avoid tripping and shutdown. In addition, the computations must be fast enough to deliver control responses with a fast tact rate of at most one millisecond on an embedded controller hardware. The goal is to deliver at least partial torque to the compressors in the presence of disturbances and thus to considerably enlarge the time until the compressors go into surge.

Mathematical research

The only control approach that can systematically deal with multiple inputs, nonlinear constraints, and economic optimization objectives simultaneously is Nonlinear Model Predictive Control (NMPC) [9]. Traditionally, black box simulation and optimization prevail in numerical methods for NMPC: An optimization solver is used repeatedly to solve a suitably discretized optimal control problem on a finite prediction horizon. The first part of the optimal control is then fed back to the system to close the feedback loop. The critical input to the optimal control problem is the current estimate of the system state at each sampling instance. In practice, this black box approach is severely limited by the usually rather long time it takes to solve one optimal control problem, which results in a high feedback delay and a possibly large lower bound on the feedback rate.

Over the last two decades, a computational paradigm shift facilitated NMPC of complex systems in the millisecond range. One way of motivating this homotopy-centered view on NMPC goes along the following line of thoughts: To allow for exploitation of the solution on the previous sampling interval, an optimization method with good hotstarting capabilities must be employed, such as Sequential Quadratic Programming (SQP). If new system measurements arrive at a fast rate, the SQP iterations will soon run on outdated information. Hence, the Real-Time Iteration [5] performs only one SQP iteration per sampling time to always incorporate the latest information. This creates the additional possibility of splitting each iteration into a comparably more costly preparation phase, which is independent of the current system state estimate, and a comparably shorter feedback phase. The preparation phase consists of system simulation and of computing its linearization. The feedback phase comprises the solution of a Quadratic Programming (QP) problem that is parametric in the current system state. Hence, the feedback delay can be drastically reduced by the preparation phase time. Further drastic increase of the feedback rate is possible on the basis of inexact SQP methods with Multi-Level Iterations [4], which exploit that the system linearization's validity usually varies on a much slower time-scale than the system trajectories to save considerable computational effort by freezing linearizations. Finally, a homotopy paradigm leads to a solution method for the solution of the parametric QP subproblems with a primal-dual active set method with fast hotstarts, the Online Active Set Strategy [7].

Implementation

The homotopy based NMPC approach was applied successfully at the Equinor site in Kollsnes. The open source software packages ACADO [8] and qpOASES [7] were employed to solve the resulting nonlinear optimization problems in real-time on embedded hardware [2, 3].

Industrial relevance and summary

The control scheme was implemented on two load commutated inverters each powering a 41.2 MW gas compressor. The Model Predictive Controller achieves sampling times faster than a millisecond and can deliver much better dynamic performance than less advanced control schemes. "During voltage dip events in the winter, the ability of the model predictive controller to provide partial torque was verified in practice," say Besselmann et al. [3].

Acknowledgments

The authors acknowledge the support by the Federal Ministry of Education and Research (BMBF project no. 05M2013, GOSSIP).

References

[1] ABB press release "ABB's Model Predictive Torque Control (MPTC) protects against gas supply interruptions." http://new.abb.com/drives/media/abbs-model-predictive-torque-control-mptc-protects-against-gas-supply-interruptions, 2016.

[2] T.J. Besselmann, S. Van de moortel, S. Almér, P. Jörg, and H.J. Ferreau, "Model predictive control in the multi-megawatt range." *IEEE Transactions on Industrial Electronics*, 63(7), 4641-4648, 2015.

[3] T.J. Besselmann, A. Cortinovis, S. Van de moortel, A.M. Ditlefsen, M. Mercangöz, H. Fretheim, P. Jörg, E. Lunde, T. Knutsen, and T.O. Stava, "Increasing the robustness of large electric driven compressor systems during voltage dips." *IEEE Transactions on Industry Applications,* 54(2), 1460-1468, 2017.

[4] H.G. Bock, M. Diehl, P. Kühl, E. Kostina, J.P. Schlöder, and L. Wirsching, "Numerical methods for efficient and fast nonlinear model predictive control." *In Assessment and future directions of nonlinear model predictive control* (pp. 163-179). Springer, Berlin, Heidelberg, 2007.

[5] H.G. Bock, M. Diehl, D.B. Leineweber, and J. Schlöder, "Efficient direct multiple shooting in nonlinear model predictive control," in *Scientific Computing in Chemical Engineering II,* F. Keil, W. Mackens, H. Voß, and J. Werther, eds., Springer-Verlag, Berlin, pp. 218–227, 1999.

[6] M. Diehl, H.G. Bock, and J.P. Schlöder. "A real-time iteration scheme for nonlinear optimization in optimal feedback control." *SIAM Journal on control and optimization,* 43(5), 1714-1736, 2005.

[7] H.J. Ferreau, C. Kirches, A. Potschka, H.G. Bock, and M. Diehl, "qpOASES: A parametric active-set algorithm for quadratic programming." *Mathematical Programming Computation,* 6(4), 327-363, 2014.

[8] B. Houska, H.J. Ferreau, and M. Diehl, "ACADO toolkit–An open-source framework for automatic control and dynamic optimization." *Optimal Control Applications and Methods,* 32(3), 298-312, 2011.

[9] D.Q. Mayne, and H. Michalska. "Receding horizon control of nonlinear systems." *Proceedings of the 27th IEEE Conference on Decision and Control.* IEEE, 1988.

Superconducting magnets at the Large Hadron Collider (LHC) at CERN

© Springer Nature Switzerland AG 2021
H. G. Bock et al. (eds.), *German Success Stories in Industrial Mathematics*,
Mathematics in Industry 35, https://doi.org/10.1007/978-3-030-81455-7_3

MATHEMATICAL MODELING, SIMULATION AND OPTIMIZATION FOR CERN'S QUENCH PROTECTION SYSTEM

A co-simulation approach for multiphysical, multiscale and multirate problem

Superconducting magnets are used to generate high magnetic fields and are employed in several applications, such as in particle accelerators to control the beam of particles that is travelling through them. The superconducting material can, under certain circumstances, quench, that is, lose its superconductivity and as a consequence get potentially highly damaged. Therefore, high energy magnets in particle accelerators such as the LHC (Large Hadron Collider) at CERN (European Organization for Nuclear Research) embed quench protection systems which mitigate the quench and avoid costly damage. These systems shall be included in quench simulations such that their performance can be assessed and possibly improved. It poses a highly multiphysical, multiscale and multirate problem, involving the coupling of different systems of equations with different properties that

are solved with different, specific simulations tools. Therefore, co-simulation techniques are employed, which allow to simulate the different systems separately by iteratively exchanging information between them. To ensure (fast) convergence of the algorithms that are used, the mathematical analysis of the convergence and the structure of the systems that are involved is of high importance. The collaboration between research institutions such as CERN and PSI (Paul Scherrer Insitut), and the Technische Universität Darmstadt under the STEAM (Simulation of Transient Effects in Accelerator Magnets) project, allowed to create a co-simulation framework, capable of simulating the quench protection system of the LHC at CERN by also understanding the mathematical foundations below the algorithms that are employed and thus ensuring their correct behavior and implementation.

IDOIA CORTES GARCIA
SEBASTIAN SCHÖPS
Computational Electromagnetics Group, TU Darmstadt

PARTNERS

LORENZO BORTOT **and** MATTHIAS MENTINK, **CERN, Switzerland**

Industrial challenge and motivation

Nowadays, low-temperature superconducting (LTS) magnets are used in high energy particle accelerators such as the Large Hadron Collider at the European Organization for Nuclear Research (CERN) in Geneva, the Paul Scherrer Institut (PSI) in Villigen or the GSI Helmholtzzentrum für Schwerionenforschung in Darmstadt.   The magnets apply electromagnetic forces on particle beams traveling through the accelerator structure.   This force is used e.g. to keep the beam focused, bend its path and to ensure its circular trajectory. The same magnet technology is also used commercially, for example in cancer treatment with facilities in Heidelberg and Marburg. If temperature, magnetic field or current in a magnet are above its critical surface then the superconducting material can quench, that is, lose its superconductivity and become resistive.   The margins for operations are tight, e.g. the critical temperature of a LTS material is about 9 K for $Nb - Ti$, and 16 K for Nb_3Sn). If the critical surface is violated, Ohmic losses will quickly heat up the magnet and thermo-electric-mechanical stress is created.   In particular, the energy density stored in the magnet is such that the peak temperature can lead to irreversible damage of the coils, compromising the integrity of the overall circuit.   The reparation of such an accident is very time consuming and costly, e.g. the busbar quench incident at CERN in 2008, and thus the study of quench propagation and protection is of high importance. In particular the design of a robust control system which quickly ramps down the magnets but does not interfere in normal operations is a challenge which can be mitigated by mathematical modelling, simulation and optimization.

The simulation of quench propagation is a problem of multiple scales and multiple time-rates, which involves the coupling of several physical phenomena solved with different equations, e.g. electromagnetic, thermal and mechanical field equations as well as electric circuits. The STEAM (Simulation of Transient Effects in Accelerator Magnets[1]) project is a cooperation between research institutions, such as CERN, PSI, and the Technische

[1] see https://cern.ch/steam

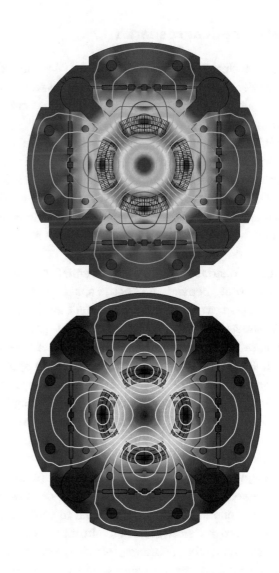

Figure 1: Superconducting magnets are used to generate high magnetic fields and are employed in several applications, such as in particle accelerators to control the beam of particles that is travelling through them.

Universität Darmstadt to develop a software platform for quench simulations that uses state of the art models and algorithms for simulation. Therefore, various models and formulations have been mathematically analyzed and implemented in software. New algorithms have been designed such that convergence can be assured. Models and algorithms are rigorously tested and complex simulation scenarios are validated by measurements.

Mathematical research

The physical phenomena involved in quench simulation are described by different systems of equations. For example, the protection circuit surrounding the superconducting magnets is modelled as a directed graph with algebraic or differential voltage-to-current relations on the branches, where the different elements are located, see Fig. 2. However, the electromagnetic field inside the magnet is described with Maxwell's equations, as this allows to establish a detailed spatially resolved information of the field description. Therefore, spatially distributed partial differential equations (PDEs) that depend on both space and time are coupled to only time dependent systems of differential algebraic equations (DAEs) [7]. To simulate the coupled system, the method of lines is used, where the PDEs are first discretized in space, to then resolve the only time dependent system in time. After the spatial discretization, coupled subsystems of DAEs are obtained, with internal degrees of freedom and inputs that depend on the solution of the other subsystems.

A design decision was to use existing (possibly proprietary) simulation tools that focus on one or several phenomenon but are not able to capture all the required physical behavior. In either case, a classical monolithic approach, which solves all the coupled subsystems together would have lead to prohibitive computation times by not exploiting the multi-rate behavior. The STEAM platform combines various simulation tools by means of a hierarchical co-simulation and the waveform relaxation technique [10]. This allows to solve the subsystems separately with different time-steps and iteratively exchange information between them so as to converge to the coupled solution.

The co-simulation of all these subsystems is organized with a waveform relaxation method [1]. Its convergence for coupled DAEs strongly depends on the index of the coupled system and the subsystems that are involved. The index of a DAE is a

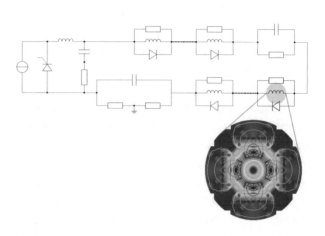

Figure 2: Superconducting accelerator circuit connecting several magnets

natural number that allows to classify them in terms of the numerical and analytical difficulties they might involve and their sensitivity towards perturbations. The index can intuitively be thought of as a measure of how many time differentiations the DAE is away from an ordinary differential equation. Lower index systems carry less difficulties.

For ordinary and differential algebraic equations of index 1, the convergence theory for the waveform relaxation method is known (see [2, 3]) and can therefore be guaranteed. In the case of index 2 circuits coupled to ordinary differential equations, convergence is ensured under some specific topological characteristics of the circuit [4]. Thus, to ensure the co-simulation scheme converges, as well as assess the possible numerical difficulties that may arise when simulating the systems, it is important to conduct an index study of the coupled DAEs [8, 9]. Based on this knowledge, the information exchange between the subsystems on the waveform relaxation scheme can be chosen to ensure their index is as low as possible.

Apart from ensuring convergence of the waveform relaxation scheme in the context of DAE analysis, the convergence can be sped up in terms of optimized Schwarz methods [5]. These study the iteration scheme as a domain decomposition method in time and use domain decomposition approaches to optimize the information exchange between the subsystems and thus reduce the number of iterations that are required to converge to the coupled solution. For the specific case of the co-simulation of LTS superconducting magnets and electric circuits, an optimized transmission condition is computed in [6], which corresponds to embedding the magnet as an inductance inside the circuit.

Implementation

While analyzing the problem and designing the numerical algorithms, the STEAM collaboration implemented a software platform for the co-simulation of transient effects in accelerator magnets and circuits at CERN. The platform was developed following the SCRUM methodology, and implemented in an open source, object-oriented programming language (Java), this is to promote a collaborative environment. The development of the code followed best coding practices, such as the object-oriented paradigm, unit tests, and periodical code reviews by experts. A the same time, this would not have been possible without profiting from cutting-edge tools for software development, such as Gitlab for versioning, Docker for code deployment and testing and SonarQube for code quality and security. Overall, this approach improved the maintainability and usability of the code, succeeding in combining the efforts of multiple developers working simultaneously on the same code.

In detail, the co-simulation platform is organized on a three-layer, scalable and expandable structure, which reflects the "divide et impera" strategy for tackling complex multi-rate and multi-scale quench simulation problems. The top layer contains the hierarchical co-simulation algorithm implementing the waveform relaxation method. The middle layer exchanges information between the models that participate on the co-simulation. The bottom layer implements a modular structure, composed by blocks called tool adapters. Each adapter implements a common interface, which acts as a "contract" between the framework and the specific tool. Once the contract is fulfilled, signals can be exchanged between the communication bus and the tool-specific models via a suitable Application Programming Interface (API), which is tool-dependent. In this way, both proprietary and in-house simulation tools can be linked with the framework on demand, by simply developing dedicated tool adapters.

The overall co-simulation process is monitored by a hierarchical state-machine algorithm for the management of the models. Depending on the status of the system under simulation, different transient phenomena may appear at different moments with different time-spans. The hierarchical algorithm embeds the knowledge about the causality relations existing between the models composing the system. Thus, the algorithm decides how the models participate in the co-simulation, and for how long. In this way, the overall computational cost is reduced.

Industrial relevance and summary

The STEAM platform has already proven its flexibility in tackling complex and various problems. It is actively used in the analysis of the LHC, its upgrade HL-LHC and for the studies of the FCC-Future Circular Collider.

In addition to CERN, also other international universities and research institutions use the software for their simulation purposes, such as INFN (National Institute for Nuclear Physics) in Italy, LBNL (Lawrence Berkeley National Laboratory) and FNAL (Fermi National Accelerator Laboratory) in the US, CIEMAT in Spain and PSI (Paul Scherrer Institute) in Switzerland. In June 2019, a two-day workshop was hold at CERN for the users of STEAM. There, an introduction into its architecture and modules was made, and assistance in the usage of STEAM was provided.

For the future of particle accelerators, the study of HTS (high temperature superconducting) magnets is becoming more and more important. Their modelling and simulation poses new mathematical challenges, that have to be tackled in the future. Here, the STEAM platform can provide an already set-up co-simulation framework, that eases the study of transient effects in HTS magnets.

Acknowledgements

The authors acknowledge the support by the Federal Ministry of Education and Research (BMBF projects no. 05M2018RDA, PASIROM and 05P18RDRB1, Diagnose and 05E15CHA, Gentner program).

References

[1] E. Lelarasmee, A. E. Ruehli, and A. L. Sangiovanni-Vincentelli, "The waveform relaxation method for time-domain analysis of large scale integrated circuits," *IEEE Trans. Comput. Aided. Des. Integrated Circ. Syst.*, vol. 1, no. 3, pp. 131–145, 1982.

[2] K. Burrage, "Parallel and sequential methods for ordinary differential equations". Oxford: Oxford University Press, 1995.

[3] A. Bartel, M. Brunk, M. Günther, and S. Schöps, "Dynamic iteration for coupled problems of electric circuits and distributed devices," *SIAM J. Sci. Comput.*, vol. 35, no. 2, pp. B315–B335, 2013.

[4] J. Pade, and C. Tischendorf, "Waveform relaxation: a convergence criterion for differential-algebraic equations," *Num. Alg.*, vol. 81, pp. 1327–1342, 2018.

[5] M. Al-Khaleel, M. J. Gander, and A. E. Ruehli, "Optimization of transmission conditions in waveform relaxation techniques for RC circuits," *SIAM J. Numer. Anal.*, vol. 52, no. 2, pp. 1076–1101, 2014.

[6] I. Cortes Garcia, S. Schöps, L. Bortot, M. Maciejewski, M. Prioli, A. M. Fernandez Navarro, B. Auchmann, and A. P. Verweij, "Optimized field/circuit coupling for the simulation of quenches in superconducting magnets," *IEEE J. Multiscale Multiphys. Comput. Tech.*, vol. 2, no. 1, pp. 97–104, 2017.

[7] S. Schöps, H. De Gersem, and T. Weiland, "Winding functions in transient magnetoquasistatic field-circuit coupled simulations," *COMPEL*, vol. 32, no. 6, pp. 2063–2083, 2013.

[8] A. Bartel, S. Baumanns, S. Schöps, "Structural analysis of electrical circuits including magneto-quasistatic devices.," APNUM **61**, pp. 1257–1270, 2011.

[9] I. Cortes Garcia, S. Schöps, C. Strohm, and C. Tischendorf, "Generalized elements for a structural analysis of circuits", Progress in Differential-Algebraic Equations II, Springer, 2020.

[10] L. Bortot, B. Auchmann, I. Cortes Garcia, A.M. Fernandez Navarro, M. Maciejewski, M. Mentink, M. Prioli, E. Ravaioli, S. Schöps, and A.P. Verweij, "STEAM: a hierarchical co-simulation framework for superconducting accelerator magnet circuits," *IEEE Trans. Appl. Super.*, vol. 28, pp. 1051–8223, 2018.

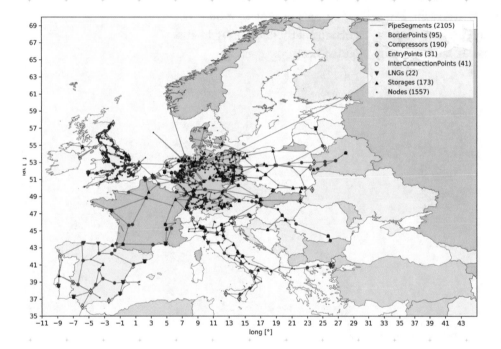

H. G. Bock et al. (eds.), *German Success Stories in Industrial Mathematics*,
Mathematics in Industry 35, https://doi.org/10.1007/978-3-030-81455-7_4

Projection-based model reduction for nonlinear gas network input-output models

Simulations of the gas network infrastructure play an important role in energy supply and the green energy transition. Especially volatilities induced by renewable energies increase the need for more transient simulations in shorter time-spans. This challenge is answered by the mathematical technology of model reduction for accelerating gas network simulations.

Practically, a software platform is developed to comprehensively test proposed model reduction methods as part of the **MathEnergy** project.

CHRISTIAN HIMPE
SARA GRUNDEL
PETER BENNER
Max Planck Institute for Dynamics of Complex Technical Systems Magdeburg

PARTNERS

PSI AG, Dortmund und Berlin

Industrial challenge and motivation

Natural gas and biogas are a principal part of Germany's, Europe's and the world's energy source. To transport the gas from the refineries or digesters to the consumers, networks of pipelines span continents, connecting supply with demand. These gas transport networks are complex in itself (see Figure 1 for a map of the European gas transport network), due to the expanse and interconnected topology. Yet, additionally, non-pipeline components, first and foremost compressor stations (see Figure 2 to get an idea of their complexity), complicate the network system furthermore.

To balance supply and demand, the network needs to be controlled with regard to operation safety, market rules, national and international regulations, and since recently, increasing volatility. In Germany, this volatility roots in the energy transition towards renewable energies: On the one hand, the fast response of gas-fired power plants can compensate short-term absence of renewable power influx, on the other hand consumers may suddenly decrease their demand due to sufficient renewable energy, and even new supplies of synthetic gas or hydrogen can materialize in case of excess renewable energy.

To handle this aggravating network control problem, gas infrastructure providers need simulations for development, long- and short-term planning as well as real-time control. For each task, many scenarios need to be simulated, to ensure safe delivery of denominations. Nowadays, for example, short-term planning of gas transport has a 24 hour horizon. Hence, ever more simulations for various unsteady uncertainties, on a large-scale network of various configurable components have to be completed in less than a day – every day.

In practice, specialized software companies, that provide the simulation and control tooling to the gas industry, like the *PSI Software AG*, have to adapt to this shortened forecasting cycle by accelerating simulations. Such a catalyst is the mathematical field of model order reduction: Gas network models, resulting in high-dimensional stiff and nonlinear differential equation systems of orders beyond 10 000, are reduced in dimensionality by model reduction,

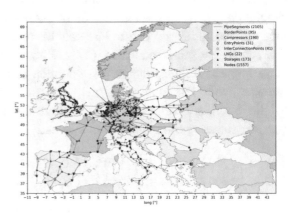

Figure 1: Visualization of a part of the European gas transport network, reconstructed from data by the SciGRID_gas project (https://gas.scigrid.de).

preserving (approximately) quantities of interest and features like multi-dimensional parametric dependencies.

Mathematical research

The basis of gas network modeling are the Euler equations for a cylindrical pipe. This system of partial differential equations describes the conservation of mass, momentum and energy. In the case of gas networks, often an isothermal variant is employed, treating temperature (energy) as constant, and due to the slow velocities the kinetic term is assumed to be close to zero and thus discarded, yielding the isothermal Euler equation system [1]:

$$\partial_t p + \frac{1}{a}\partial_x q = 0,$$
$$\partial_t q + a\partial_x p + agp\gamma^{-1}\partial_x h(x) = -\frac{\lambda\gamma}{2da}\frac{q|q|}{p}. \tag{1}$$

Connecting pipes to a network introduces further conservation properties at the junctions, equivalent to Kirchhoff's laws for electrical circuits, leading to a partial differential-algebraic equation system; for details see [4].

Since the length of the pipes exceeds their diameter by far, a one-dimensional spatial discretization of

(1) is acceptable. Yet, due to the hyperbolicity of the Euler equations, this spatial discretization needs to match the sought temporal resolution of the simulation. Then, for the special case of gas network models, the algebraic constraints can be resolved analytically by index reduction [4]. Altogether, a large implicit ordinary differential equation system remains, with the inputs $\begin{pmatrix} s_p & d_q \end{pmatrix}^\mathsf{T}$ corresponding to the boundary values, and an output function $\begin{pmatrix} s_q & d_p \end{pmatrix}^\mathsf{T}$ filtering the quantities of interest:

$$
\begin{pmatrix} E_p & 0 \\ 0 & I_q \end{pmatrix} \begin{pmatrix} \dot{p}(t) \\ \dot{q}(t) \end{pmatrix} = \begin{pmatrix} 0 & A_p \\ A_q & 0 \end{pmatrix} \begin{pmatrix} p(t) \\ q(t) \end{pmatrix} + \begin{pmatrix} 0 & B_p \\ B_q & 0 \end{pmatrix} \begin{pmatrix} s_p(t) \\ d_q(t) \end{pmatrix}
$$
$$
+ \begin{pmatrix} 0 \\ f_q(p(t), q(t), s_p(t)) \end{pmatrix}, \qquad (2)
$$
$$
\begin{pmatrix} s_q(t) \\ d_p(t) \end{pmatrix} = \begin{pmatrix} 0 & C_q \\ C_p & 0 \end{pmatrix} \begin{pmatrix} p(t) \\ q(t) \end{pmatrix}.
$$

Here, nonlinearity f_q encodes the friction and compressor nonlinearities.

Now, this nonlinear input-output system has to be simulated for many scenarios of inputs, for example in an optimization process, or a bulk test. Yet, repeatedly simulating this system is time consuming, since the state variables p and q are high-dimensional for expansive networks.

To make repeated simulations for manifold scenarios feasible, the field of model reduction provides methods to reduce the dimensionality of a system's state-space algorithmically, and thus accelerates its simulation. Generally, model reduction is at the intersection of the disciplines of numerical mathematics, scientific computing and computational science & engineering. Yet, specifically, a major branch of model reduction is concerned with systems such as (2). This system-theoretic model reduction builds on linear system theory, which investigates linear input-output systems, given by:

$$
E\dot{x}(t) = Ax(t) + Bu(t),
$$
$$
y(t) = Cx(t). \qquad (3)
$$

The principle of model reduction for such systems is dimensionality reduction of x, such that the mapping from inputs u to outputs y is as accurate as possible, but can be evaluated much faster than the original model. In the context of gas network models, this

Figure 2: Natural gas compressor station in Werne, Germany, run by Open Grid Europe.

means reducing the dimensionality of p and q, while preserving the map from boundary values (available supply and ordered demand variables, s_p and d_q) to quantities of interest (sensors s_q and d_p), but explicitly not approximating flow behavior in every pipeline meter.

The general approach to model reduction is projection-based. Mathematically, this means that the state trajectory is confined to a subspace of the full phase space. To this end, a projection-based model reduction method computes a reduction operator V and a reconstruction operator U, such that a reduced state x_r is constructed by $x_r := Vx : x \approx Ux_r$, $VU = I$, with $x_r(t)$ having a much smaller dimension than $x(t)$. The practical application of projection-based model reduction to (2) is illustrated in Figure 3.

The gas network input-output system (2) is nonlinear, and a linearization would distort the essential friction effect and eliminate certain compressor models. Hence, the linear system-theoretic model reduction methods are not applicable. At the same time, fully nonlinear system-theoretic model reduction approaches are currently not computationally feasible for the considered class of gas network models.

Alternatively, the model reduction sub-project of the *MathEnergy* project, funded by the German Federal Ministry for Economic Affairs and Energy

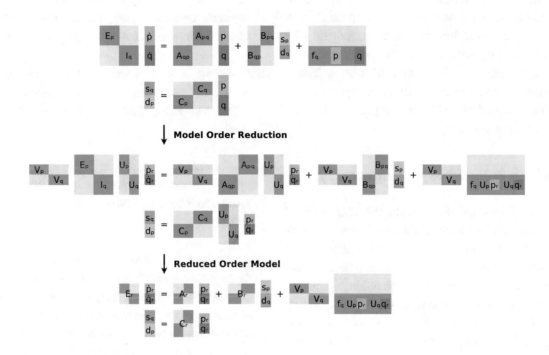

Figure 3: An illustration of projection-based model reduction for the nonlinear gas network input-output systems (2).

(BMWi), investigates a data-driven generalization of linear system-theoretic methods, which were already successfully applied in the field of neuroscience to accelerate simulations for the recovery of brain connectivity from measurements [5]. In these methods, the nonlinearity is approximated by trajectory data of systematically perturbed steady-state simulations. This class of data-driven projection-based system-theoretic model reduction methods can handle the nonlinear gas network models, when used in a structured manner to preserve the separate pressure and mass-flux variables p and q in the reduction process.

Nonetheless, the hyperbolic character of the Euler equations makes matters more complicated. A promising candidate among the methods under investigation is the so-called *structured empirical cross-Gramian-based dominant subspaces* method, derived from [2], which, additionally to preserving structure and approximating nonlinearity, seems advantageous for hyperbolic systems [3], compared to similar methods. However, without analytic proof of superiority for any of the tested methods,

an exhaustive heuristic comparison needs to be conducted to inspire confidence with the industrial partner.

Implementation

The deliverable of the *MathEnergy* project is a software library tackling the challenges of the evolving energy grids, to which the model reduction sub-project contributes a software platform enabling the test, comparison and evaluation of model reduction algorithms. Furthermore, beyond model reduction methods, also different mathematical gas network models and time-stepping solvers can be tested, due to a fully modular design. The overall requirements and modeling details are discussed in regular meetings with specialists from the *PSI Software AG*, who also provided sample simulation results to validate our own solvers. With this comparison platform, it will soon be possible for a given target network, as well as a selected model and solver to determine the most suitable model reduction method. This promotes an industrial adaptation of model reduction not only based on a seemingly appropriate method proposed in a publication, but resulting from data-supported test series and in competition with alternative methods. Lastly, to maximize the reach of our software platform, we plan to release it under an open-source license and compatible with a proprietary and open-source software stack.

Industrial relevance and summary

In this contribution, we illustrate the challenge of accelerating transient simulations of gas network models, which features various mathematical complexities. For the sake of exposition, we have omitted some of the difficulties, such as additional parameter dependency of the system's vector field. Nonetheless, model reduction techniques for faster gas network simulations are available, in the form of data-driven system-theoretic methods. But due to the complexities of real-life applications, like gas transport, which elude theoretical results on suitability, these methods need to be comprehensively tested heuristically. For this purpose the model reduction sub-project of *MathEnergy* is providing the software platform, which is currently tested in an academic environment, but with real data, and will serve as a prototypical implementation.

Acknowledgement

Supported by the German Federal Ministry for Economic Affairs and Energy (BMWi), in the joint project: "MathEnergy – Mathematical Key Technologies for Evolving Energy Grids", sub-project: Model Order Reduction (Grant number: 0324019**B**). The authors thank the SciGRID_gas project at the DLR Oldenburg for preparing and providing the graphic in Figure 1.

References

[1] P. Benner, S. Grundel, C. Himpe, C. Huck, T. Streubel, and C. Tischendorf. Gas network benchmark models. In *Applications of Differential Algebraic Equations: Examples and Benchmarks*, Differential-Algebraic Equation Forum, pages 171–197. Springer Cham, 2018. doi:10.1007/11221_2018_5.

[2] P. Benner and C. Himpe. Cross-Gramian-based dominant subspaces. *Adv. Comput. Math.*, 45(5):2533–2553, 2019. doi:10.1007/s10444-019-09724-7.

[3] S. Grundel, C. Himpe, and J. Saak. On empirical system Gramians. *Proc. Appl. Math. Mech.*, 19(1):e201900006, 2019. doi:10.1002/PAMM.201900006.

[4] S. Grundel, L. Jansen, N. Hornung, T. Clees, C. Tischendorf, and P. Benner. Model order reduction of differential algebraic equations arising from the simulation of gas transport networks. In *Progress in Differential-Algebraic Equations*, Differential-Algebraic Equations Forum, pages 183–205. Springer Berlin Heidelberg, 2014. doi:10.1007/978-3-662-44926-4_9.

[5] C. Himpe. *Combined State and Parameter Reduction for Nonlinear Systems with an Application in Neuroscience*. PhD thesis, Westfälische Wilhelms-Universität Münster, 2017. Sierke Verlag Göttingen, ISBN 9783868448818. doi:10.14626/9783868448818.

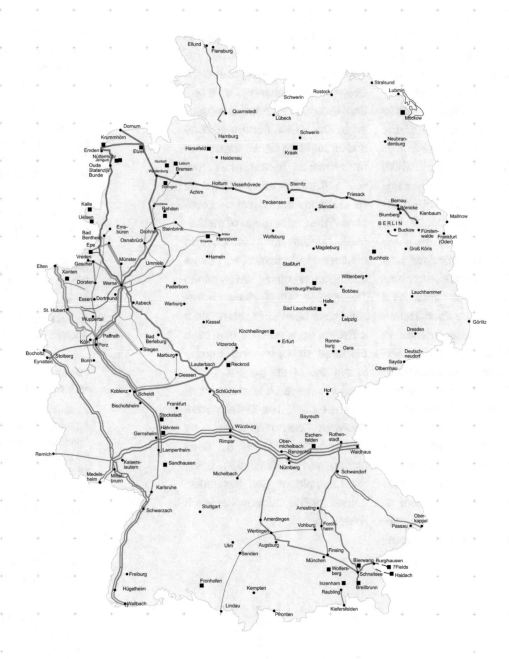

Network systems of Open Grid Europe. The network model is split into three parts: H-gas north (purple), H-gas south (red), and L-gas (green). (Source: Open Grid Europe.)

CAPACITY EVALUATION FOR LARGE-SCALE GAS NETWORKS

A discrete-continuous model for optimal transport of gas

Natural gas is important for the energy turnaround in many countries like in Germany, where it serves as a bridging energy towards a fossil-free energy supply in the future. About 20% of the total German energy demand is provided by natural gas, which is transported through a complex pipeline network with a total length of about 30000 km, and the efficient use of the given transport infrastructure for natural gas is of political, economic, and societal importance. As a consequence of the liberalization of the European gas market in the last decades, gas trading and transport have been decoupled. This has led to new challenges for gas transport companies, and mathematical optimization is perfectly suited for tackling many of these challenges. However, the underlying mathematical problems are by far too hard to be solved by today's general-purpose software so that novel mathematical theory and algorithms are needed. The industrial research project 'ForNe: Research Cooperation Network Optimization' has been initiated and funded by Open Grid Europe in 2009 and brought together experts in mathematical optimization from seven German universities and research institutes, which cover almost the entire range of mathematical optimization: integer and nonlinear optimization as well as optimization under uncertainty. The mathematical research results have been put together in a software package that has been delivered to Open Grid Europe at the end of the project. Moreover, the research is still continuing – e.g., in the Collaborative Research Center/Transregio 154 „Mathematical Modelling, Simulation and Optimization using the Example of Gas Networks" funded by the German Research Foundation.

MARTIN SCHMIDT
Trier University, Department of Mathematics

BENJAMIN HILLER
atesio GmbH

THORSTEN KOCH
(a) TU Berlin, Chair for Software and Algorithms for Discrete Optimization, (b) Zuse Institute Berlin

MARC E. PFETSCH
Technische Universität Darmstadt, AG Optimierung

BJÖRN GEISSLER
Adams Consult GmbH & Co. KG, Büttelborn

RENÉ HENRION
Weierstrass Institute for Applied Analysis and Stochastics

IMKE JOORMANN
TU Braunschweig, Institute for Mathematical Optimization

ALEXANDER MARTIN
(a) Friedrich-Alexander-Universität Erlangen-Nürnberg
(b) Fraunhofer Institute for Integrated Circuits IIS

ANTONIO MORSI
Adams Consult GmbH & Co. KG, Büttelborn

WERNER RÖMISCH
Humboldt-University Berlin, Institute of Mathematics

LARS SCHEWE
University of Edinburgh, School of Mathematics

RÜDIGER SCHULTZ
Universität Duisburg-Essen, Faculty of Mathematics

MARC C. STEINBACH
Leibniz Universität Hannover, Institute of Applied Mathematics

The original version of this chapter was revised: The authors names and affiliation have been corrected in the xml file. The correction to this chapter is available at https://doi.org/10.1007/978-3-030-81455-7_27

Industrial challenge and motivation

Since the liberalization of the gas market in Europe by the EU Commission from 1998 on, trade and transport of natural gas are decoupled and have to be conducted by separate companies. Today, the European transmission system operators (TSOs) usually operate under the so-called entry-exit system [1, 5]. A brief explanation of this system is as follows. First, the TSOs publish technical capacities for all the points in their network at which gas can be supplied or withdrawn. Second, gas traders can then book a capacity right at these nodes up to the previously published technical capacity of the node. This booking serves as a mid- to long-term capacity-right contract. Third, on a day-to-day basis, these traders can nominate a certain amount up to their booked capacity and only have to ensure that all their nominations are in balance, i.e., the total amount of supplied gas has to match the total amount of withdrawn gas. Fourth and lastly, the TSO has to transport the actual nomination.

The main complexity lies in the fact that the TSOs have to be able to transport any balanced nomination that is in compliance with the respective bookings. Thus, the feasibility of transport needs, in principle, to be checked for an infinite number of possible nominations. This is in contrast to the goal of the TSOs to publish technical capacities for possible bookings that are as large as possible. Consequently, for an efficient usage of resources and long-term planning, it is important to compute the capacities of the network.

There are two major challenges for TSOs like Open Grid Europe: First, the usage of capacities by the gas traders is uncertain, i.e., it is unknown which nominations will arise. Second, deciding whether a nomination can be operated through the network involves both integer and highly nonlinear aspects. The task is thus properly modeled as a stochastic mixed-integer nonlinear optimization problem. Since these problems need to be solved on complex and large-scale transport networks, there was a strong need for new mathematical techniques for solving these challenging optimization problems.

Thus, Open Grid Europe formed an industrial research project that included seven universities and

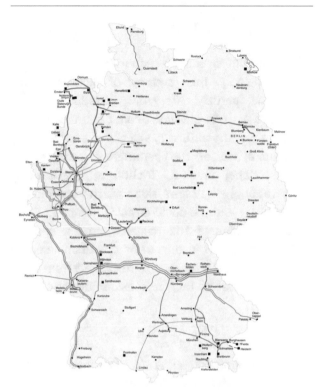

Figure 1: The natural gas transport network of Open Grid Europe. The mathematical optimization model for the optimal transport of gas through such a network combines almost all challenges of modern optimization: integer controls, differential equations, highly nonlinear models of physics and engineering, as well as uncertain data.

research institutions in Germany, covering the range of integer, stochastic, and nonlinear optimization.

Mathematical research

Since by signing a booking contract the TSO guarantees to transport any possible load flow situation (that complies to this booking), checking the feasibility of nominations lies at the heart of the research project. A rigorous mathematical modeling of gas transport needs to consider different elements of a gas transport network. Pipes are outnumbering all other devices in these networks. The flow of natural gas through a pipe can be modeled by using the Euler equations of compressible fluids in cylindrical pipes:

$$\frac{\partial \rho}{\partial t} + \frac{1}{A}\frac{\partial q}{\partial x} = 0,$$

$$\frac{1}{A}\frac{\partial q}{\partial t} + \frac{\partial p}{\partial x} + \frac{1}{A}\frac{\partial (qv)}{\partial x} = -\lambda(q)\frac{|v|v}{2D}\rho - g\rho h'.$$

Together with a suitable equation of state, e.g., $p = \rho R_s T z$, this system of nonlinear and hyperbolic partial differential equations (PDEs) couples the main physical quantities mass flow q, pressure p, temperature T, velocity v, and density ρ in dependence of the parameters diameter D, cross-sectional area A, friction λ, gravitational acceleration g, the pipe's slope h', the specific gas constant R_s, and the compressibility factor z. Since the main focus of the research project was on planning problems instead of on operational control of the network, it is reasonable to abstract from time-dependent effects, i.e., to consider the stationary variant

$$\frac{\partial q}{\partial x} = 0, \qquad \frac{\partial p}{\partial x} + \frac{1}{A}\frac{\partial (qv)}{\partial x} = -\lambda(q)\frac{|v|v}{2D}\rho - g\rho h'$$

of the above mentioned PDEs. To control the flow through the network and its pressure, other network elements like compressors or control valves need to be operated. The respective models usually are of mixed-integer type and the mathematical problem of cost-optimal gas transport thus translates to a mixed-integer nonlinear optimization problem with ordinary differential equations. The day-to-day practice in planning departments of many TSOs is that the feasibility of a specific nomination is checked by using simulation software. Here, the user needs to configure the controllable elements of the transport network and the simulation tool computes a physical state based on this control, for which feasibility can be checked afterward. In the case of infeasibility, the controls need to be adapted manually and the process is repeated. Here, mathematical optimization comes into play because it allows to automatize this process. Our solution approach for validating the feasibility of a nomination is sketched in the lower block of the flow chart given in Figure 2. Since the original mixed-integer nonlinear problem is by far too hard to be solved on real-world networks, the solution process is split up into two different components. First, a variety of physically simplified models based on mixed-integer

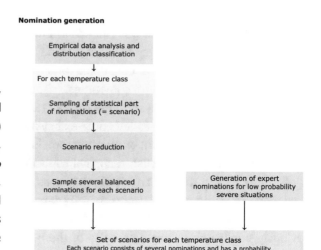

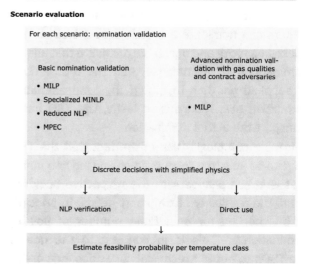

Figure 2: Flow chart of the solution approach for validating the feasibility of a booking contract.

linear optimization, mathematical programming with equilibrium constraints, mixed-integer nonlinear optimization, and purely continuous optimization are solved to guess reasonable integer controls of the network. Afterward, the feasibility of these controls are verified by using a detailed nonlinear model that represents physical and technical requirements very precisely.

On top of that, verifying the feasibility of a booking has to account for all possible nominations that might occur. This requirement needs to be relaxed for practice and is replaced by an approach that guarantees the feasibility of a nomination with a high probability. The respective nomination generation approach is sketched in the upper block of Figure 2. First, by analyzing historical gas flow data from several years, we estimate probability distributions modeling the demands at the exits. This information and the capacity contracts in question are used to sample the space of possible nominations, which are then checked for feasibility by the techniques described above. Finally, the results of thousands of samples are combined and analyzed to give an estimate for the probability of validity of the offered capacities. For the details, we refer to [2, 3, 4, 6].

Implementation

The research project "ForNe: Research Cooperation Network Optimization"[1] was initiated in 2009 by Open Grid Europe, Germany's largest gas network operator, and conducted jointly by seven research institutes and universities: Zuse Institute Berlin (Mathematical Optimization), Friedrich-Alexander Universität Erlangen-Nürnberg (Discrete Optimization), Leibniz Universität Hannover (Institut für Angewandte Mathematik), Universität Duisburg-Essen (Fakultät für Mathematik), Technische Universität Darmstadt (Fachbereich Mathematik), Humboldt-Universität zu Berlin (Institut für Mathematik), and the Weierstraß-Institut für Angewandte Analysis und Stochastik Berlin.

The success of the project was, in particular, due to regular meetings of the scientific partners with the industrial partner Open Grid Europe that took place multiple times per year. During these meetings, Open Grid Europe not only provided us with the relevant data but the discussions also gave insights into the structure of the studied problems that later helped to develop efficient solution techniques.

Based on these measures, the project contributed a new methodology for this problem, including innovative solution approaches for mixed-integer nonlinear optimization problems under uncertainty. It also triggered the research project "Untersuchung der technischen Kapazität von Gasnetzen", which was funded by the German Federal Ministry of

Economic Affairs and Energy (Bundesministerium für Wirtschaft und Energie – BMWi) during the years 2009–2012. Moreover, it has been supported by the Bundesnetzagentur (BNetzA).

Industrial relevance and summary

A comprehensive software implementing our methods was evaluated on real-world large-scale networks and delivered to Open Grid Europe. In the course of the project, ten PhD theses and more than 20 research articles have been published. The developed methods are comprehensively described in the book "Evaluating gas network capacities" [3], which received the 2016 EURO Excellence in Practice Award. As part of the above mentioned BMWi project, elaborate instance format descriptions and a library of gas network instances were published at `http://gaslib.zib.de`; see also [8]. Moreover, the research results also led to the text book [7] that already served as a basis in applied mathematical optimization lectures at German universities.

The research of this project is also continued in the Collaborative Research Center/Transregio 154 "Mathematical Modelling, Simulation and Optimization using the Example of Gas Networks" funded by the German Research Foundation, which started in 2014 and which is still going on.

Acknowledgements

The work presented in this article has been supported by the German Federal Ministry for Economic Affairs and Energy (BMWi) in the project "Untersuchung der technischen Kapazität von Gasnetzen" (fund number 0328006A), and by the DFG funded SFB Transregio 154 "Mathematical Modelling, Simulation and Optimization using the Example of Gas Networks".

References

[1] A. Fügenschuh, B. Geißler, R. Gollmer, C. Hayn, R. Henrion, B. Hiller, J. Humpola, T. Koch, T. Lehmann, A. Martin, R. Mirkov, A. Morsi, J. Rövekamp, L. Schewe, M. Schmidt, R. Schultz, R. Schwarz, J. Schweiger, C. Stangl, M. C. Steinbach, and B. M.

---

[1] `https://www.zib.de/projects/forne-research-cooperation-network-optimization`

Willert. Mathematical optimization for challenging network planning problems in unbundled liberalized gas markets. *Energy Systems*, 5(3):449–473, 2014.

[2] B. Hiller, T. Koch, L. Schewe, R. Schwarz, and J. Schweiger. A system to evaluate gas network capacities: Concepts and implementation. *European Journal of Operational Research*, 270(3):797–808, 2018.

[3] T. Koch, B. Hiller, M. E. Pfetsch, and L. Schewe, editors. *Evaluating Gas Network Capacities*. MOS-SIAM Series on Optimization. SIAM, 2015.

[4] T. Koch, H. Leövey, R. Mirkov, W. Römisch, and I. Wegner-Specht. Szenariogenerierung zur Modellierung der stochastischen Ausspeiselasten in einem Gastransportnetz. In *Optimierung in der Energiewirtschaft*, VDI-Berichte 2157, pages 115–125. VDI-Verlag, Düsseldorf, 2011.

[5] A. Martin, B. Geißler, C. Hayn, A. Morsi, L. Schewe, B. Hiller, J. Humpola, T. Koch, T. Lehmann, R. Schwarz, J. Schweiger, M. Pfetsch, M. Schmidt, M. Steinbach, B. Willert, and R. Schultz. Optimierung Technischer Kapazitäten in Gasnetzen. In *Optimierung in der Energiewirtschaft*, VDI-Berichte 2157, pages 105–114, 2011.

[6] M. E. Pfetsch, A. Fügenschuh, B. Geißler, N. Geißler, R. Gollmer, B. Hiller, J. Humpola, T. Koch, T. Lehmann, A. Martin, A. Morsi, J. Rövekamp, L. Schewe, M. Schmidt, R. Schultz, R. Schwarz, J. Schweiger, C. Stangl, M. C. Steinbach, S. Vigerske, and B. M. Willert. Validation of nominations in gas network optimization: models, methods, and solutions. *Optimization Methods and Software*, 30(1):15–53, 2015.

[7] L. Schewe and M. Schmidt. *Optimierung von Versorgungsnetzen. Mathematische Modellierung und Lösungstechniken*. Springer Spektrum, Berlin, Heidelberg, 2019.

[8] M. Schmidt, D. Aßmann, R. Burlacu, J. Humpola, I. Joormann, N. Kanelakis, T. Koch, D. Oucherif, M. E. Pfetsch, L. Schewe, R. Schwarz, and M. Sirvent.

GasLib-a library of gas network instances. *Data*, 2(4), 2017.

Funktional principle PURITY SCANNER ADVANCED

1. Material feed
2. Transport system
3. X-ray inspection
4. Optical inspection
5. Sorting unit
6. Rejected material
7. Clean material
8. Infrared camera
9. Color camera

© Springer Nature Switzerland AG 2021
H. G. Bock et al. (eds.), *German Success Stories in Industrial Mathematics*,
Mathematics in Industry 35, https://doi.org/10.1007/978-3-030-81455-7_6

PURITY ASSESSMENT OF PELLETS USING DEEP LEARNING

A study of deep neural networks for quality control

Plastic pellets – an unknown, but key ingredient for virtually anything made of plastic such as extra-high voltage cables, medical technology or advanced material. This modern technology, however, requires ever increasing purity of plastic granulate. To match this, the SIKORA AG, Bremen, developed a *PURITY SCANNER* system which ensures high material quality by continuous examination and by sorting out defect material. This high-throughput system is destined to be integrated in various production steps of the polymer industry and could provide a crucial step forward towards pure material production.

To even go further on the way to guaranteed purity in plastic manufacturing, one important feature of the *PURITY SCANNER* system still needs to be developed. Currently, the system does separate contaminated pellets from clean pellets using camera systems, but does not allow a continuous feedback system about the specific types of faults detected during its operation. Such a system would be of great interest since a real-time feedback to the operators of the production would allow the producers to act fast, correct the sources of contamination and finally reduce costly production rejects.

The task of detecting faulty pellets can be considered as an image analysis task, which is well-suited for recently developed deep learning algorithms. As a collaboration of the Center for Industrial Mathematics, University of Bremen, and the R&D department of the SIKORA AG, Bremen, we conducted a study to assess the potential of neural network based vision systems for real-time pellet detection. In this report, we describe our initial study, next steps towards deep learning-based detection and the mathematical background required to successfully develop problem adapted solutions. To summarize, this study is an example of how world-class engineering departments like R&D from the SIKORA AG benefit from close interaction with mathematical research groups like the Center for Industrial Mathematics.

JENS BEHRMANN
MAXIMILIAN SCHMIDT
JANNIK WILDNER
PETER MAASS
Center for Industrial Mathematics, University of Bremen

PARTNERS

DR. SEBASTIAN SCHMALE | SIKORA AG, Bremen

Industrial challenge and motivation

The purity of advanced materials, as they are used in medical technology, semiconductor manufacturing, aerospace industries and extra-high voltage cables is a decisive characteristic for the quality of the final product. Therefore, the purity of the plastic granulate is of the utmost priority for the polymer industry. For example in underwater high-voltage cables, metal specks in plastic coating can cause severe damages and are hard to repair underwater. With the *PURITY SCANNER*, SIKORA provides an unrivaled, user oriented system for 100 % online inspection with automatic sorting of plastic pellets at all process levels. The system intelligently combines X-ray technology with a flexible optical system, see Figure 1. In this combination, the X-ray camera assures the detection of metallic impurities inside the pellet as well as on its surface. Thus, all pellets are reliably inspected. Discolorations in transparent or on translucent and colored raw materials are identified by a color and/or black and white camera as optical faults. Contaminated pellets are automatically sorted out.

To get an impression of the sheer number of pellets being analyzed during operation, lets consider the following example: the *PURITY SCANNER* can process 6000 pellets per second. Each optical camera produces image rows with a frequency of 32 kHz, which results in 7 high-resolution images (8192 × 4096 pixels) per second. Thus, a standard system with 2 cameras can produce up to $2 \cdot 7 \cdot 6000 \approx 84\,000$ Pellet images per second, which have to be checked and sorted out.

While the *PURITY SCANNER* by SIKORA ensures high material quality, it does currently not allow for an online interpretation system of images of pellet defects. Sorted out pellets can only be analyzed manually afterwards, resulting potentially in delays of crucial feedback to the operating system. For example, an online system could provide an instant warning to the production system about types of contamination. This in turn would allow a faster response and efficient handling.

The goal of this study was to assess whether deep learning allows to detect contamination in pellets in an online fashion. This could be used in the future

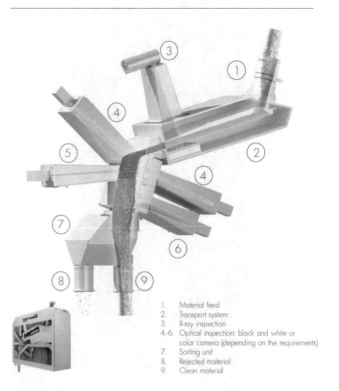

1. Material feed
2. Transport system
3. X-ray inspection
4.-6. Optical inspection: black and white or color camera (depending on the requirements)
7. Sorting unit
8. Rejected material
9. Clean material

Figure 1: Schematic description of the workflow of the *PURITY SCANNER*. In particular, the system provides flexible camera options for all requirements stemming from high-throughput production.

as an assistance system that enables the *PURITY SCANNER* to provide continuous feedback to the production process.

Mathematical background and methodology

The study focused on deep learning algorithms for quality assessment using black and white high-resolution images, see Figure 2 (top) for an example. To tackle the task of detecting pellets, we used a combination of classical image analysis approaches and deep learning. In particular, the classical image analysis pipeline was introduced by SIKORA and used as ground-truth data for the detection task, see Figure 2 (bottom).

First, we start with the mathematical background that builds the basis for our algorithms. Deep learning is usually implemented by neural networks,

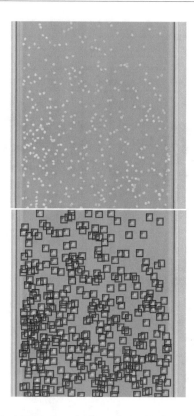

Figure 2: (Top) Pellets from high-resolution black and white camera from the *PURITY SCANNER*. (Bottom) Detected pellets by developed deep learning algorithm.

which are composition of parametrized affine and non-linear activation functions. In particular, a neural network $F_\theta : \mathbb{R}^{d_1} \to \mathbb{R}^{d_2}$ can be defined as

$$F_\theta = f_\theta^L \circ \cdots \circ f_\theta^1,$$

$$\text{where} \quad f_\theta^i(x^i) = \phi^i \left(A^i x^i + b^i \right), \quad i \in [L]$$

denotes a layer. Further, $A^i \in \mathbb{R}^{m_i \times n_i}$ is called a linear layer, $b^i \in \mathbb{R}^m$ a bias and ϕ^i an elementwise non-linear function. In our case, the linear mappings A^i are implemented using multi-channel discrete convolutions and inputs x are 2-dimensional images. Given training data $\mathcal{T} = \{x^{(i)}, y^{(i)}\}_{i=1}^N$, we minimize the empirical risk $L_\mathcal{T}$ with respect to parameters $\theta \in \mathbb{R}^p$

$$\theta^* \in \underset{\theta \in \Theta}{\arg\min} \, L_\mathcal{T}(F_\theta) = \underset{\theta \in \Theta}{\arg\min} \, \frac{1}{N} \sum_{i=1}^N \ell\big(F_\theta, (x^{(i)}, y^{(i)})\big).$$

In our task, we perform both classification of pellet contamination and pellet detection. Classification can be described statistically as maximum likelihood

estimation with categorically distributed labels y. In this setting, the loss function ℓ is the cross-entropy loss and the neural network F_θ outputs a distribution

$$p_\theta(Y = \mathbf{k} \mid X = x) = \mathsf{softmax}(F_\theta(x)_k)$$
$$= \frac{\exp(F_\theta(x)_k^T)}{\sum_{k'=1}^C \exp(F_\theta(x)_{k'})},$$

where $Y = \mathbf{k}$ denotes that the target is from class k. The pellet detection task can be described as a regression problem, where the coordinates of the boxes are to be approximated. Following the standard approach in deep learning based object detection, we employ a smoothed ℓ_1 loss for regression [5, 7]. Given these fundamental concepts, we employ three recently introduced methods for object detection:

- SSD: Single Shot MultiBox Detector [5]

- Faster R-CNN [7]

- FCOS - Fully Convolutional One-Stage Object Detection [8].

All these approaches use convolutional neural networks (linear layer A^i implemented by multi-channel discrete convolution) to extract features from input images x. The SSD and Faster R-CNN approach rely on anchor boxes, where the corners of the bounding box are given by some default values and the regression task is to predict the offset. FCOS, on the other hand, directly predicts these bounding boxes using regression. For the classification task, we used a residual network [4].

To summarize our used methodology relies crucially on neural networks, an area where our basic research strongly focuses on. Examples of our works include invertible residual networks [2], robustness and interpretability of neural networks [3] and data-based modeling in inverse problems [1].

Numerical results

In this section, we will briefly summarize our numerical result on two test dataset provided by SIKORA. First, we started with an evaluation of the detection models, see Table 1. As an evaluation metric we used the mean intersection over union (mIoU) measure, see [5, 7, 8]. While all methods perform similarly on the validation sets, there is a large gap in terms of runtime. Considering the fast processing of pellets in the *PURITY SCANNER*, the runtime is of crucial importance for real-time image processing. To conclude this pre-study on the detection task would favor the Faster R-CNN approach. Furthermore, future work should consider ways to combine the classical image analysis workflow with deep learning.

In addition to detecting pellets, we studied the classification task of distinguishing pellets with black specks from clean pellets. One challenging aspect of this task, is the severely misbalanced proportion of clean pellets to contaminated pellets. To tackle this problem, we employ a weighting (w_0 for clean pellets, w_1 for bad pellets) based on the relative abundance as

$$w_0 = \frac{n_1}{n_0 + n_1} \approx 0,003 \quad \text{and} \quad w_1 = \frac{n_0}{n_0 + n_1} \approx 0,997.$$

For training we used a residual network [4] with 18 layers, implemented in *PyTorch* [6]. Besides a quantitative evaluation using area-under-curve scores, we emphasized our analysis on predictions where ground truth and predictions differed most, see Figure 3. Most strikingly, the neural networks were able to detect contamination in pellets that were wrongly labeled as clean. Thus, in a follow-up study we plan to use an interactive approach, where the ground truth data can be corrected by looking at the strongest differences between model prediction and label.

Industrial relevance and transfer of mathematics to industry

This study on detecting plastic pellets in a real-time and reliable way is highly relevant in many industrial productions steps. While the *PURITY SCANNER* by SIKORA already provides an offline system to separate certain contaminated pellets from clean pellets, our study focused on enhancing this system. In particular, we analyzed how deep learning approaches can detect pellets in a real-time setting, which would allow continuous feedback to the production. When a certain fault is detected, the operators would receive immediate feedback and could act accordingly. The dataset used

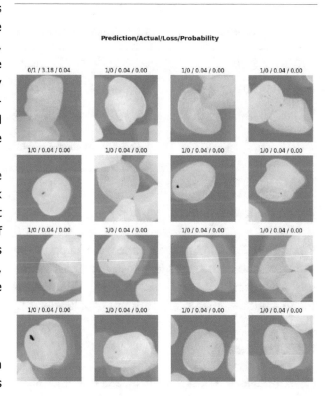

Figure 3: Visualization of pellets, showing both contaminated and clean pellets. The chosen examples correspond to the largest mismatch between model prediction and label. However, the mismatch is rather due to incorrect labels. Hence, the predictions can be used to correct ground truth labels to further improve the used dataset.

for this study was only intended to get a first impression on how deep learning approaches work in such an industrial system. While the results were promising, most real-world scenario will be more challenging. We believe, however, that the combined mathematically expertise of the Center for Industrial Mathematics (ZeTeM, Bremen) and the engineering skills from SIKORA is a great starting ground for further explorations. As an example, we started to transfer the knowledge from research into the R&D department of SIKORA by implementing training sessions at the ZeTeM, sharing source code and an intense interaction. In the future, we hope to continue and intensify this process to further strengthen the transfer of advanced mathematics into industrial practice.

Model	mIoU Train	mIoU Val	Runtime [sec]	Images per second
SSD	0.72413	0.70487	0.15819	6.3
Faster R-CNN	0.70919	0.70987	0.03686	27.1
FCOS	0.72002	0.71903	0.15414	6.5

Table 1: Numerical results of different detection methods.

Acknowledgments

Maximilian Schmidt acknowledges the support by the Deutsche Forschungsgemeinschaft (DFG) within the framework of GRK 2224/1 "π^3: Parameter Identification – Analysis, Algorithms, Applications ".

References

[1] S. Arridge, P. Maass, O. Öktem, and C.-B. Schönlieb. Solving inverse problems using data-driven models. *Acta Numerica*, 28:1–174, 2019.

[2] J. Behrmann, W. Grathwohl, R. T. Q. Chen, D. Duvenaud, and J.-H. Jacobsen. Invertible residual networks. In *Proceedings of the 36th International Conference on Machine Learning (ICML)*, pages 573–582, 2019.

[3] C. Etmann, S. Lunz, P. Maass, and C. Schoenlieb. On the connection between adversarial robustness and saliency map interpretability. In *Proceedings of the 36th International Conference on Machine Learning (ICML)*, pages 1823–1832, 2019.

[4] K. He, X. Zhang, S. Ren, and J. Sun. Deep residual learning for image recognition. In *2016 IEEE Conference on Computer Vision and Pattern Recognition (CVPR)*, pages 770–778, 2016.

[5] W. Liu, D. Anguelov, D. Erhan, C. Szegedy, S. Reed, C.-Y. Fu, and A. C. Berg. SSD: single shot multibox detector. In *European Conference on Computer Vision (ECCV)*, 2016.

[6] A. Paszke, S. Gross, F. Massa, A. Lerer, J. Bradbury, G. Chanan, T. Killeen, Z. Lin, N. Gimelshein, L. Antiga, A. Desmaison, A. Kopf, E. Yang, Z. DeVito, M. Raison, A. Tejani, S. Chilamkurthy, B. Steiner, L. Fang, J. Bai, and S. Chintala. Pytorch: An imperative style, high-performance deep learning library. In *Advances in Neural Information Processing Systems 32 (NeurIPS)*, pages 8024–8035. 2019.

[7] S. Ren, K. He, R. Girshick, and J. Sun. Faster r-cnn: Towards real-time object detection with region proposal networks. In *Advances in Neural Information Processing Systems 28 (NIPS)*, pages 91–99. 2015.

[8] Z. Tian, C. Shen, H. Chen, and T. He. FCOS: Fully convolutional one-stage object detection. In *Proc. Int. Conf. Computer Vision (ICCV)*, 2019.

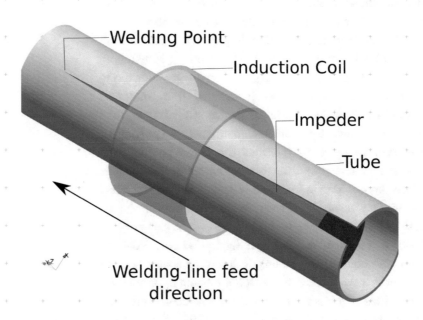

Schematic visualization of a high-frequency induction welding process.

© Springer Nature Switzerland AG 2021
H. G. Bock et al. (eds.), *German Success Stories in Industrial Mathematics*,
Mathematics in Industry 35, https://doi.org/10.1007/978-3-030-81455-7_7

MODELLING AND SIMULATION OF HIGH-FREQUENCY INDUCTION WELDING

Modelling and optimization of a process chain for steel production and manufacturing

PRERANA DAS
EFD Induction, Skien, Norway

DIETMAR HÖMBERG
Weierstrass Institute, Berlin

Almost all manufacturing sectors, from construction through transport to consumer goods, are largely based on the utilisation of steels. Steel products often compete favourably with alternative material solutions in cost efficiency and life cycle analyses. The last fifteen years have seen the development of ever more refined high-strength and multiphase steels with purpose designed chemical compositions allowing for significant weight reduction, e.g., in automotive industry. The production of these modern steel grades needs a precise process control, since there is only a narrow process window available in which the desired physical properties are defined. In combination with component walls getting thinner and thinner, these new steels make also new demands on a more precise process control in metal manufacturing processes, such as welding and hardening.

Improved and optimised process control requires quantitative mathematical modelling, simulation and optimisation of the complex thermal cycles and thermal gradients experienced by the processed material. Such models require an understanding of the behaviour of the materials from a materials science and phase transformations perspective. Unfortunately, it is almost impossible for companies to find graduates combining deep knowledge in materials science with expertise in mathematical modelling, simulation and optimisation.

To fill this gap five partners from steel production (Outokumpu, SSAB) and steel manufacturing (EFD Induction), from materials science (University of Oulu) and applied mathematics (WIAS) established the European Industrial Doctorate program on "Mathematics and Materials Science for Steel Production and Manufacturing (MIMESIS)", where eight PhD projects were jointly carried out, providing a unique interdisciplinary and inter-sectorial training opportunity.

The research was focused on three major topics along the process chain for steel production and manufacturing, secondary metallurgy in the ladle, phase transformations during steel production and steel manufacturing processes based on induction heating. As a specific case study we present research results related to high-frequency induction welding of steel tubes.

PARTNERS

EFD INDUCTION, Skien, Norway

Industrial challenges and motivation

High-frequency induction welding is widely used, especially in the production of superior quality oil and gas pipes and structural tubes. A steel strip is cold-formed into a tubular shape in a continuous roll forming mill. The strip edges are electromagnetically heated and joined mechanically by pushing the strip edges against each other to form the longitudinally welded tube.

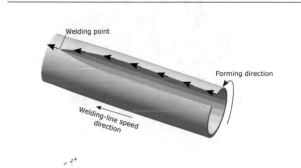

Figure 1: Local velocity to represent feed velocity and forming.

The welded joint as seen in the transverse cross-section of a welded tube, is a very narrow zone compared to the tube diameter. The strip edges are heated to almost melting temperature and are pushed against each other in the viscoplastic state to form the welded joint where crystallographic texture and microstructural changes appear.

The electromagnetic heating of the tube is analogous to transformer theory. The coil is the primary current source, the strip is where the current is induced and the impeder acts as a magnetic core. The entire setup, coil current and frequency determine the amount of the induced current in the tube.

High-frequency alternating current is supplied to the induction coil. This induces eddy currents in the strip under the coil. The induced current can follow the principal paths indicated in the figure above to complete the circuit. Along the strip edges, it can flow downstream from the coil towards the welding point or away from the coil in the upstream direction. At any strip cross-section, the current can follow a path either along the outer circumference or the inner circumference. The goal of high-frequency induction welding is to maximise the current density in the strip edge downstream, towards the welding point.

The relative positioning of the strip, induction coil and impeder is very important to obtain an efficient heating process. The geometric shape of the opening between the strip edges is usually a Vee-shape. Sometimes it is distorted by spring-back due to the mechanical forming of the strip. This also affects the current distribution.

Further important process parameters are the coil current and welder frequency. For better understanding of the complex interactions between the above parameters, numerical simulation is an indispensable tool.

Mathematical research

In [1] we present the first comprehensive simulation approach for high-frequency induction welding in 3D. Its main novelties are a new analytic expression for the space-dependent velocity of tubes accounting for arbitrary Vee-angle and spring-back and a stabilisation strategy, which allows us to consider realistic welding-line speeds.

The mathematical model comprises a harmonic vector potential formulation of the Maxwell equations and a quasi-static, convection dominated heat equation coupled through the joule heat term and nonlinear constitutive relations.

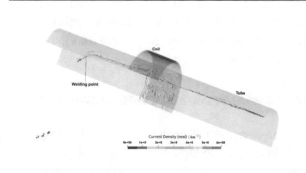

Figure 2: Current path in the tube.

An important effect that needs to be accounted for is the computation of velocity in the strip. After the welding point is reached the velocity has only a component in the feeding direction, while before the velocity varies locally with non-vanishing radial and angular components. To obtain the correct temperatures especially close to the strip edge it is crucial to use the correct locally varying velocity for the simulation. Figure 1 shows the resulting local velocity vectors for selected points on the strip opening for a spring-back tube opening. Instead of a constant velocity solely in y-direction one can see that now the velocity follows the contour of the opening, for details, we refer to [1].

The heat equation is discretised by linear nodal finite elements. To account for high speeds somewhere in the range of 40 m/min to 200 m/min the Streamline Upwind Petrov Galerkin (SUPG) method is utilised. The discretisation of Maxwell's equations is done with Nédélec elements of lowest order. The magnetisation depends both on the temperature and the magnetic field. For fixed temperature this nonlinearity is resolved numerically based on an averaging approach [2]. The coupled system is iteratively decoupled and solved using a fixed point iteration.

Figure 2 is a simulation result of current density distribution in the strip edge. It shows current concentration both in the downstream and upstream directions with a maximum close to the weld point.

Examples of temperature distribution in the welded strip are shown in Figure 3 for two different Vee-openings and a spring-back distorted opening. The strip edges are heated to very high temperatures because of Joule heating from eddy current concentration. The velocity function incorporates the mechanical forming of the strip into a tube in addition to the welding-line velocity.

Industrial relevance and summary

A three-dimensional model has been developed for high-frequency induction welding. It is a non-linearly coupled system of Maxwell's electromagnetic equation and the heat equation. The results show a temperature distribution in the strip edges that develops as expected from previous studies and visual observations of the process. The strip length also decides the amount of induced current that goes to the welding point. A wider Vee-angle results in a wider heat affected zone. Increasing the frequency

reduces the width of the heat affected zone. It is also shown that for the thinner wall the hour glass shape of heat affected zone is less pronounced. These results are in line with what is expected.

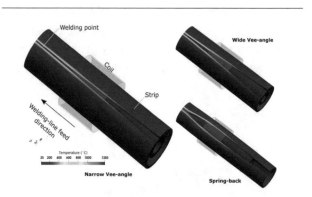

Figure 3: Temperature distribution in the tube for different openings.

This new three-dimensional simulation tool provides a basis for an optimisation of the design of the welder, especially with respect to the dimensioning of induction coil, impeder and the configuration of these relative to the steel strip. Future work will include the study of the mechanics of the material squeeze-out when the strip edges are joined together after heating.

Acknowledgements

The authors acknowledge the support by the European Union's Horizon 2020 research and innovation programme under the Marie Skłodowska-Curie grant agreement No. 675715 (MIMESIS).

References

[1] J. I. Asperheim, P. Das, B. Grande, D. Hömberg and T. Petzold, *Numerical simulation of high-frequency induction welding in longitudinal welded tubes*. WIAS Preprint no 2600, Berlin 2019.

[2] D. Hömberg, Q. Liu, J. Montalvo-Urquizo, D. Nadolski, T. Petzold, A. Schmidt, A. Schulz, *Simulation of multi-frequency-induction-hardening including phase transition and mechanical effects*. Finite Element Analysis Design **200** (2016), 86–100.

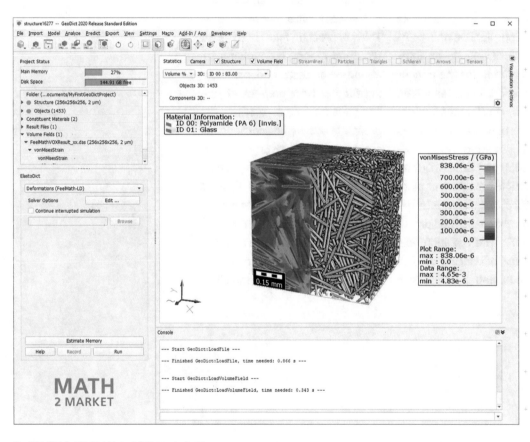

GeoDict GUI for Digital Material Characterization

© Springer Nature Switzerland AG 2021
H. G. Bock et al. (eds.), *German Success Stories in Industrial Mathematics*,
Mathematics in Industry 35, https://doi.org/10.1007/978-3-030-81455-7_8

DIGITAL MATERIAL CHARACTERIZATION

Design of New Composites and Components

Fiber reinforced plastics have a high stiffness to weight ratio and can be produced in a cost-effective way on a mass production scale by injection or compression molding. Therefore, this type of material plays an important role for producing lightweight components. We developed an integrative simulation for the dimensioning of short fiber reinforced components, which takes into account the production process as well as the resulting locally varying material properties. During the production process the plastic is injected or molded into the component shape at medium to high pressure. The resulting flow processes are influencing the fiber orientation and by this, also the mechanical properties significantly. The gap between the process simulation and the structure design is bridged by Digital Material Characterization, i.e. full field simulations on digital twins of the composite microstructure. This approach allows to save time and money by avoiding trial-and-error tests and prototyping for the development of new components. The core of this approach is our highly efficient micro-mechanical solver FeelMath. Our spin-off company Math2Market, which was founded in 2012 and employs over 30 people in 2019, is distributing FeelMath worldwide as part of the software GeoDict (www.geodict.com) for Digital Material Characterisation.

MATTHIAS KABEL
JONATHAN KÖBLER
HEIKO ANDRÄ
**Fraunhofer Institute ITWM,
Kaiserslautern**

PARTNERS

ERIC GLATT, **Math2Market GmbH, Kaiserslautern**

Industrial challenge and motivation

To fully exhaust a material's lightweight potential, accurate models for the material properties are of paramount importance. Due to the large differences in the mechanical properties of the constituents and the complex geometric shapes associated to reinforcements of composite materials, analytical models possess only a restricted predictive potential, in particular when nonlinear effects enter the stage. Digital material characterization bridges the gap between the process simulation (e.g. injection molding) and the structure design, by providing microscopic composite material properties for component simulations and make it possible to answer the following challenging questions:

- Does the new composite material improve my component?

- Does it combine sufficient stiffness with low weight?

- Do I understand the microscopic material properties of the composite needed for the simulation of macroscopic properties of the component?

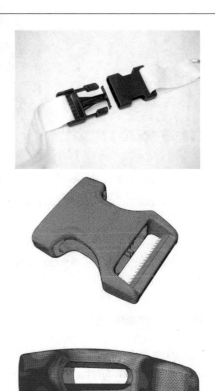

Figure 1: Component simulation with Abaqus

Implementation of the initiative

The development of the micromechanics solver FeelMath started in a series of publicly funded research and industrial projects at the Fraunhofer ITWM. To better understand which problems are industrially relevant, a long-term collaboration for the integration of FeelMath in the digital material laboratory GeoDict was initiated between Math2Market and Fraunhofer ITWM. Math2Market provides the technical framework and effects the first level support for our joint customers. Fraunhofer ITWM enhances the software according to the needs of the industry.

To bridge the gap between process simulation and structure design, the digital material characterization had to be automatized, and interfaces to commercial simulation tools (e.g. Moldflow, Abaqus) had to be implemented. More precisely, the following steps are performed for the component simulation (see Figure 1) taking into account locally varying fiber orientations [2, 6]

1. Offline phase

    - Generate representative volume elements (RVEs) for at least 15 different fiber orientations (see Figure 2) [4].

    - Perform microscale simulation for the RVEs and use model order reduction methods to obtain effective nonlinear mechanical models.

2. Online phase

    - Map the fiber orientations of the injection simulation (e.g. by Moldflow) onto the Finite Element Mesh for the component simulation (e.g. with Abaqus)

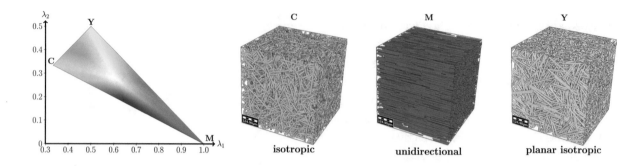

Figure 2: Fiber orientation triangle with two degrees of freedom, that constitute the essentially different fiber orientations

- "Interpolate" between the effective nonlinear mechanical models (see Figure 3 and Figure 4).

Mathematical research

In the recent years, the FFT-based homogenization method of Moulinec-Suquet [3] has emerged as a powerful tool for the computation of effective mechanical properties of micro-structured heterogeneous materials at small strains. To validate experimental data or to calibrate constituent material parameters genuinely mixed loading conditions are necessary to simulate for example uniaxial tensile tests and force driven tests.

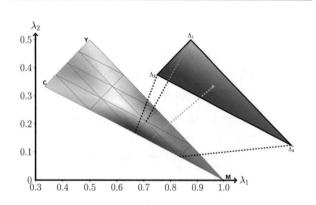

Figure 3: Interpolation concept: 1. Find adjacent orientation nodes in the fiber orientation triangle

FFT-based homogenization only has the capability to

solve for the periodic displacement field which solves the static equilibrium equation with given constant exterior strain E by

$$\text{div } \sigma(\varepsilon) = 0, \quad \varepsilon = E + \frac{\nabla u + (\nabla u)^T}{2},$$

To extend the Lippmann-Schwinger equation that stands behind the FFT-based homogenization method to mixed strain/stress averages it turned out to be useful to introduce idempotent symmetric projectors $\mathbb{P}$ and $\mathbb{Q} = \mathbb{I} - \mathbb{P}$ which encode the strain resp. stress components of the prescribed average strain/stress

$$\mathbb{P} : \langle \varepsilon \rangle_Y = E,$$

$$\mathbb{Q} : \langle \sigma(\varepsilon) \rangle_Y = \Sigma,$$

where $\langle \cdot \rangle$ denotes the volume average with respect to the RVE Y. Then a necessary condition for the prescribed strain E and prescribed stress Σ is

$$\mathbb{P} : \Sigma = 0, \quad \mathbb{Q} : E = 0,$$

which only means that it is not possible to prescribe strain and stress in the same direction. A sufficient condition for solvability is the existence of a constant strain field ε^0 with

$$\mathbb{P} : \varepsilon^0 = E, \quad \mathbb{Q} : \langle \sigma(\varepsilon^0) \rangle_Y = \Sigma.$$

The use of the projectors $\mathbb{P}$ and $\mathbb{Q} = \mathbb{I} - \mathbb{P}$ also allows to prescribe averages with respect to an independent coordinate system, e.g.

- Pure strain boundary conditions $\mathbb{P} = \mathbb{I}$

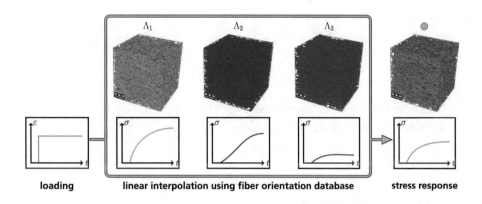

Λ_1 Λ_2 Λ_3

loading   linear interpolation using fiber orientation database   stress response

Figure 4: Interpolation concept: 1. Compute stresses in adjacent orientation nodes and interpolate stress

- Strain driven tensile test in x-direction

$$\mathbb{P} = e_1 \otimes e_1 \otimes e_1 \otimes e_1$$

- Strain driven tensile test in 45°-direction

$$\mathbb{P} = f_1 \otimes f_1 \otimes f_1 \otimes f_1 \text{ with } f_1 = (e_1 + e_2)/\sqrt{2}.$$

The solutions of the static equilibrium equation with mixed prescribed averages are critical points of the energy

$$f(\varepsilon) = \int_Y \frac{\sigma(\varepsilon) : \varepsilon}{2} - \Sigma : \varepsilon \, dx$$

subject to the constraint $\mathbb{P} : \langle \varepsilon \rangle_Y = E$. A forward Euler discretization of the negative gradient flow of f is the Lippmann-Schwinger equation for mixed boundary conditions [1]

$$\varepsilon + \left[\Gamma^0 + \mathbb{M} : \mathbb{Q} : \langle \cdot \rangle_Y \right] : (\sigma(\varepsilon) - \mathbb{C}^0 : \varepsilon)$$
$$= E + \mathbb{M} : \left[\Sigma - \mathbb{Q} : \mathbb{C}^0 : E \right]$$

with

$$\mathbb{M} = (\mathbb{Q} : \mathbb{C}_0 : \mathbb{Q})^\dagger$$

being the Moore-Penrose pseudoinverse.

Industrial relevance and summary

As part of the project, FeelMath has been integrated into GeoDict to solve nonlinear thermomechanical problems at small or large deformations using a highly innovative approach, based on the Fast Fourier Transform (FFT) and a staggered grid discretization [5]. By working directly on 3D images, the tedious mesh generation step inherent to classical Finite Element-based approaches thus can be avoided. In addition, material cards for component simulations that take into account the local changing properties of the composite material (e.g. fiber orientation, fiber volume fraction), can be generated in a fully automatic manner by using the Python interface of GeoDict. By this, the possibility is given to the product developers of our joint customers to exploit the advantages of state of the art composites without changing their workflow or increasing the simulation time during the design process.

References

[1] M. Kabel, S. Fliegener, and M. Schneider. Mixed boundary conditions for FFT-based homogenization at finite strains. *Computational Mechanics*, 57(2):193–210, 2016.

[2] J. Köbler, M. Schneider, F. Ospald, H. Andrä, and R. Müller. Fiber orientation interpolation for the multiscale analysis of short fiber reinforced composite parts. *Computational Mechanics*, 61(6):729–750, Jun 2018.

[3] H. Moulinec and P. Suquet. A fast numerical method for computing the linear and nonlinear mechanical properties of composites. *Comptes rendus de l'Académie des sciences. Série II, Mécanique, physique, chimie, astronomie*, 318(11):1417–1423, 1994.

[4] M. Schneider. The sequential addition and migration method to generate representative volume elements for the homogenization of short fiber reinforced plastics. *Computational Mechanics*, 59(2):247–263, Feb 2017.

[5] M. Schneider, F. Ospald, and M. Kabel. Computational homogenization of elasticity on a staggered grid. *International Journal for Numerical Methods in Engineering*, 105(9):693–720, 2016.

[6] J. Spahn, H. Andrä, M. Kabel, and R. Müller. A multiscale approach for modeling progressive damage of composite materials using fast fourier transforms. *Computer Methods in Applied Mechanics and Engineering*, 268:871–883, 2014.

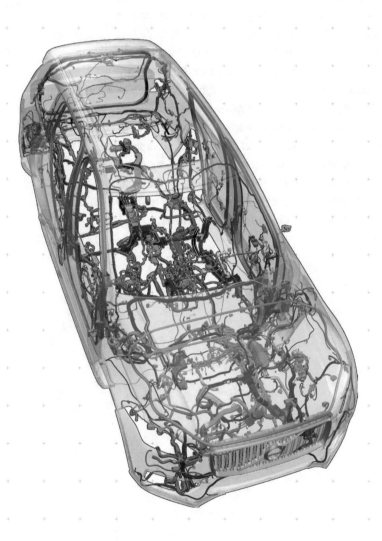

VIRTUAL PRODUCT DEVELOPMENT AND DIGITAL VALIDATION IN AUTOMOTIVE INDUSTRY

Interactive simulation of flexible cables

Engineers working with software tools for computer aided design (CAD), digital mock-up and virtual assembly need to handle rigid as well as flexible parts at interactive rates. Slender flexible parts like cables and hoses are important for the functionality of complex mechatronic machinery like cars, trucks or aircraft. The theory of *Cosserat rods* provides the proper framework to perform physically correct simulations of large spatial deformations of such structures. By combining ideas from the *discrete differential geometry of framed curves* with the variational framework of Lagrangian mechanics one may construct *discrete Cosserat rod models* that behave qualitatively correct even for coarse discretizations, provide very fast computational performance at moderate accuracy, and thus are suitable for interactive simulation. This geometry based discretization approach for flexible 1D structures has industrial applications in design, digital mock-up and digital validation, as illustrated by some application examples from automotive industry.

JOACHIM LINN
FABIO SCHNEIDER
KLAUS DRESSLER
Fraunhofer Institute ITWM, Kaiserslautern

OLIVER HERMANNS
fleXstructures GmbH

PARTNERS

FLEXSTRUCTURES GMBH, **Kaiserslautern**
IPS AB, **Göteborg, Sweden**

Industrial challenge and motivation

Standard software tools currently used in industry for CAD, digital mock-up and virtual assembly can only handle *rigid* geometries. This strongly limits the range of applications, as in many systems essential functional parts are flexible and experience larger deformation during assembly operation. Therefore there is an increasing demand for a realistic, yet easy-to-use simulation of *large deformations* of *slender flexible structures*, preferably in real time (i.e.: *at interactive rates*). Typical examples of such structures from automotive industry are tubes, hoses, single cables, or wiring harnesses collecting many cables within a compound structure.

The theory of *Cosserat rods* [1] provides a framework for structural models suitable for physically correct simulations of large deformation of slender flexible objects by *stretching*, *bending* and *twisting*. In computational mechanics, Cosserat rod models are usually discretized via nonlinear finite elements (FE) [8]. Due to their algorithmic and algebraic complexity, nonlinear FE models are technically complicated and in general computationally far too demanding to perform sufficiently fast simulations that are compatible with rendering at $25\,Hz$ (at least). This level of computational performance is required for simulations that need to run simultaneously to an interactive modification of the boundary conditions by the user, either via the graphical user interface of a desktop computer, or via a data glove (or similar devices) within a virtual reality (VR) environment. Therefore, if one aims at such interactive applications, the development of a different approach is required. The kinematics of Cosserat rods is closely related to the *differential geometry of framed curves* [4], with the differential invariants of rod configurations corresponding to the strain measures of the mechanical theory [1]. We utilize ideas from the *discrete* differential geometry of framed curves [2] to construct the *discrete kinematics* of Cosserat rod models in a way that preserves the essential geometric properties independent of the coarseness of the discretization. We model Cosserat rods within Euler's variational framework [7] utilized in *Lagrangian mechanics* in terms of the kinetic and elastic energy of the

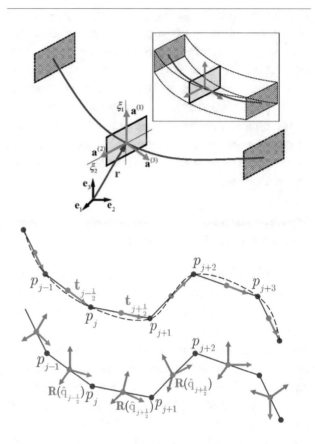

Figure 1: *Left:* Centerline $\mathbf{r}(s)$ and attached moving frame $R(s) = \mathbf{a}^{(k)}(s) \otimes \mathbf{e}_k$ of a Cosserat rod. *Right:* Polygonal arc approximating a smooth regular geometric curve $\mathcal{C}$: The vertices $p_j \in \mathcal{C}$ located at positions $\mathbf{r}_j \in \mathbb{E}^3$ define edges $[p_{j-1}, p_j]$ of length $\ell_{j-1/2}$, with edge centered tangent vectors $\mathbf{t}_{j-1/2} = (\mathbf{r}_j - \mathbf{r}_{j-1})/\ell_{j-1/2}$ of unit length. A *discrete Cosserat curve* [13, 16] consists of a polygonal arc with edge centered quaternions $\hat{q}_{j-1/2} \in S^3$ representing orthonormal frames.

rod [3, 9, 10, 11, 12]. We derive the discrete elastic energy function by an approach which we denote as *geometric finite differences* and preserves essential geometric properties. Due to the geometric discretization of the strain measures, discrete rod models constructed according to our *discrete Lagrangian mechanics* approach behave qualitatively correct even for very coarse discretizations. This in turn can be utilized to achieve high computational performance at still good accuracy.

Our discrete formulation of geometrically exact rods turns out to be particularly useful for a seamless integration into a CAE software environment like *IPS Cable Simulation*. As the models and algorithms are formulated in terms of elementary concepts of computational geometry, one can achieve the computational performance necessary for a true interaction of the user with the software in real time, which is a key feature in practical applications. We illustrate this aspect by presenting some typical application examples of simulations of cables performed in automotive industry for the qualification of product design and assembly and for digital validation.

Mathematical research and experimental mechanics

Below we briefly summarize a few facts about our discrete rod model as recently described in [6, 14, 13, 16] and refer to these articles for details, including notation.

Stable static equilibrium configurations of an elastic rod correspond to minima of its stored energy function $W^{(el)} = \int_0^L ds\, \mathcal{V}^{(el)}(\mathbf{\Gamma}, \mathbf{K})$, with *elastic energy density* $\mathcal{V}^{(el)}(\mathbf{\Gamma}, \mathbf{K}) = \frac{1}{2}\langle \Delta\mathbf{\Gamma}, C_\Gamma \cdot \Delta\mathbf{\Gamma}\rangle + \frac{1}{2}\langle \Delta\mathbf{K}, C_K \cdot \Delta\mathbf{K}\rangle$ depending on the *effective stiffness parameters* $C_\Gamma := \mathrm{diag}([GA_1], [GA_2], [EA])$ and $C_K := \mathrm{diag}([EI_1], [EI_2], [GJ])$ of the local cross section. The energy density $\mathcal{V}^{(el)}$ is a quadratic form of the deviations $\Delta\mathbf{K} = \mathbf{K}(s) - \mathbf{K}_0(s)$ and $\Delta\mathbf{\Gamma}(s) = \mathbf{\Gamma}(s) - \mathbf{\Gamma}_0$ of the material curvature $\mathbf{K} = 2\hat{q}^* \circ \partial_s\hat{q}$ and material tangent vector $\mathbf{\Gamma} = \hat{q}^* \circ \partial_s\mathbf{r} \circ \hat{q}$ from their reference values $\mathbf{K}_0(s)$ and $\mathbf{\Gamma}_0 \equiv \mathbf{e}_3$. Our approach of *geometric finite differences* results in a *mimetic discretization* of the geometry of Cosserat rod configurations by constructing a discrete differential geometry framework for framed curves [16].

It can be shown that *discrete curvatures* $\mathbf{K}_j := 2\log(\hat{q}^*_{j-1/2} \circ \hat{q}_{j+1/2})/\bar{h}_j$ and *discrete material tangent vectors* $\mathbf{\Gamma}_{j-1/2} := \hat{q}^*_{j-1/2} \circ (\mathbf{r}_j - \mathbf{r}_{j-1})/h_{j-1/2} \circ \hat{q}_{j-1/2}$ constitute a *complete set of invariants* that determines the spatial configuration of a discrete Cosserat curve uniquely up to rigid motions in Euclidian space. As this statement holds for arbitrarily coarse discretizations, the *qualitative* behaviour of a discrete elastic Cosserat rod model constructed by this approach remains correct independent of its discretization.

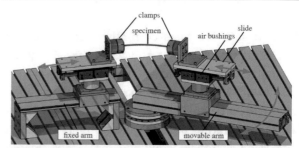
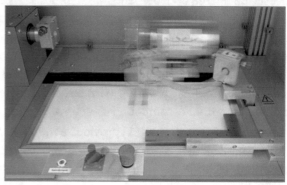

Figure 2: Test rig to measure the stiffness of cables in the case of *pure bending* [5] (*left*), and bending stiffness measurements with the *MeSOMICS* machine designed by ITWM [15] (*right*).

Experimental mechanics for Cosserat rod modeling

The application of discrete Cosserat rod models in industrial practice requires a proper knowledge of the effective stiffness properties of the slender cable or hose structure to be simulated. Motivated by customer demands, own experimental work to determine the necessary parameters was started already several years ago. In particular for bending stiffness measurements, the state of the art was found to be insuffiencient for probing the domain of larger bending curvatures. Therefore own developments of a variety of novel experimental set-ups were begun, like the two approaches displayed in Fig. 2. While for ideally elastic structures these procedures can be used for direct measurements, more complex composite structures like real cables and hoses display disctinct *inelastic* behaviour, such that the task to determine the effective bending stiffness of a cable or hose that matches the behaviour of the real structure in an optimal way implies the usage of mathematical methods for

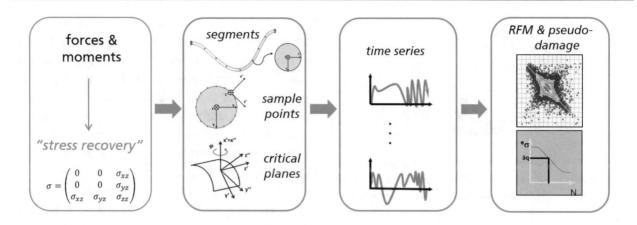

Figure 3: Essential steps of the comparative load data analysis: stress recovery, critical plane approach, rainflow counting and pseudo-damage calculation.

model based parameter identification.

Simulation-based load data analysis for cables and hoses

A recent application combining cable simulation with *numerical fatigue analysis* has been presented in [17, 18]. For both quasi-static as well as dynamic simulations, a so-called *comparative load data analysis* can be performed. The basic analysis scheme consists of several steps, as sketched in Figure 3: From the simulation, one obtains time series of cross section forces and moments along a cable or hose. These time series will be used to compute accumulated *pseudo-damage* values on the cable surface. We refer to [17, 18] and the literature cited therin for further details on the intermediate algorithmical steps of the computational procedure, consisting of *stress recovery*, the approach of *critical planes*, *rainflow counting* and *pseudo-damage calculation*, as well as typical application examples of this method.

Implementation and application examples

The software package *IPS Cable Simulation* utilizes a discrete Cosserat rod model, augmented by numerous productivity features that are useful for a variety of applications in industry, like enhanced CAD, digital mock-up, or the simulation of assembly (or disassembly) processes on desktop computers as well as in virtual reality (VR) environments. The implemented rod model has been enhanced w.r.t. a variety of different elements, such that external constraints induced by various types of clips, restricting spatial motions of one to six of the local rigid body d.o.f. of the rod, frictionless contact interaction with rigid geometries in the environment, or self contact and contact with other rods can be handled efficiently.

The user can interact with the discrete model of a cable or hose e.g. by grabbing one of its ends with the mouse or more sophisticated VR devices, or likewise a clip attached at some intermediate location, and change its position and orientation in space interactively. These changes are captured by the user interface and translated by the software into a sequential change of boundary conditions. This sequence is then passed to the simulation model, which computes corresponding deformation sequences by solving the mechanical equilibrium equations sufficiently fast for rendering deformed configuration at interactive rates (i.e.: $25\,Hz$ or faster). The seamless integration of the discrete Cosserat rod model into the IPS software is supported by its formulation in terms of elementary geometric quantities (i.e.: vertex positions and orthonormal frames) that can be handled very efficiently by the computational geometry methods

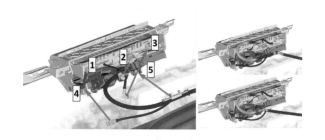

Figure 4: *Left:* Assembly sequence of various clips and connectors to be mounted at head-up display. *Right:* Variation of the position and the type of a clip to lower the level of strain in the cable.

and algorithms already implemented in IPS for different applications. As an example, frames translated along polygonal paths in space are fundamental objects in rigid body path planning, which historically has been one of the core capabilities of the IPS software.

Typical application examples from automotive industry

Below we present some examples, all taken from joint projects with AUDI AG, to illustrate the typical usage of our discrete rod models integrated in the IPS software in automotive industry (see also [13]). Fig. 4 shows some screenshots taken from the simulation of the assembly sequence of various clips and connectors to be mounted at a head up display. The numbers indicate the assembly sequence, and the arrows point out the direction of the movement of the various clips and plugs towards their final mounting positions. The engineer working with the simulation software would typically grab individual clips, plugs or connectors with the mouse and perform the assembly operation for all on the virtual model sequentially, just as it would later be done by the worker in the real process. In this way, the engineer can validate if it is possible to assemble all cables without mutual crossing, and if the cable lengths are sufficient to avoid extensional straining and sharp bending at clips. If not, the length of a cable can be changed interactively to its optimal length. Also, the function of various clip types as well as their positioning can be checked and varied to avoid infavourable configurations, as shown in the

two pictures to the right of Fig. 4.

Fig. 5 displays another issue discovered and solved during the digital validation of the functionality of the same head up display: The originally suggested design of one of the plastic parts in the kinematic mechanism was done in a compact way to assure its mechanical stability with a minimal amount of plastic material, with a rather sharp rectangular kink at its upper right corner. The picture to the right of Fig. 5 shows the final design solution: The vertical edge at the respective corner was bent upwards and elongated, such that the sharp right angle is eliminated, and the edge functions as a rail on which the cable can slide in a stable way during the forward and backward operation of the kinematic mechanism.

The simulation of the whole system with the physically correct deformation of flexible cables in contact with the rigid geometries in its environment provided the essential insights leading to an improved design. With IPS the functionality of the whole system of the headup display with all cables connected can be simulated and validated digitally in an early design phase. Traditionally, one would manufacture hardware prototypes of the simulated design to perform a physical validation. The possibility of simulation with physically correct models allows for functional validation without physical prototypes.

Industrial relevance and summary

Research and development work aiming at simulation models that are suitable for an interactive handling of slender flexible structures for assembly simulation was started in late 2004 with a team of four people at ITWM in Kaiserslautern (T. Stephan and J. Linn) and FCC in Göteborg (T. Hermansson and R. Bohlin).

The first commercially available version of *IPS Cable Simulation* was presented to industry at the Hannover fair in 2009. Based on first commercial success, two spin-off companies — *fleXstructures GmbH* in Kaiserslautern and *IPS AB* in Göteborg — were founded in 2012, which provide professional services in sales, marketing, training, deployment and engineering projects, as well as

Figure 5: Redesign of a plastic part in the kinematic mechanism of the same head-up display. *Left:* Former design with a sharp corner, where the cable may get hooked, which might cause damage. *Right:* Improved design, where the cable can slide along reproducibly always in the same stable way, such that potential damage due to hook-up or clamping of the cable is avoided.

professional software development and maintenance for industrial customers on the markets in Europe, Asia and North America. Today more than 60 people are working in the IPS teams in Kaiserslautern and Göteborg, and in industrial companies more than one thousand users are working with *IPS Cable Simulation*. Initially IPS software has been used at OEMs in automotive industry. The range of users recently spread out to cable and hose manufacturers and system suppliers. Current development aims at spreading the technology to aviation industry and electronics industry.

Mathematical modeling and simulation technology, paired with solid competences in computational and experimental mechanics, and excellent skills in software engineering needed to implement a software product of high usability have been the key factors for success of *IPS Cable Simulation* in both scientific research as well as widespread application in industry.

References

[1] Antman, S.S.: *Nonlinear Problems of Elasticity*, Springer (2005).

[2] Bergou, M., Wardetzky, M., Robinson, S., Audoly, B., Grinspun, E.: Discrete Elastic Rods, *ACM Transaction on Graphics (SIGGRAPH)*, Vol. **27**(3), pp. 63:1–63:12 (2008).

[3] Bobenko, A.I., Suris, Yu.B.: Discrete time Lagrangian mechanics on Lie groups, with an application on the Lagrange top, *Comm. Math. Phys.* Vol. **204**, pp. 147–188 (1999).

[4] do Carmo, M.P.: *Differential Geometry of Curves and Surfaces*, Prentice–Hall (1976).

[5] Dörlich, V., Linn, J., Diebels, S.: Flexible beam-like structures—experimental investigation and modeling of cables. In: Altenbach, H., et al. (eds.) Advances in Mechanics of Materials and Structural Analysis. Advanced Structured Materials, vol. **80**, pp. 27–46, Springer (2018).

[6] Dörlich, V., Hermansson, T., Linn, J.: Localized Helix Configurations of Discrete Cosserat Rods. In Kecskeméthy A., Geu Flores F. (eds.): Multibody Dynamics 2019. ECCOMAS 2019. Computational Methods in Applied Sciences, vol 53, pp. 191-198, Springer, Cham (2019)

[7] L. Euler: *Methodus inveniendi lineas curvas maximi minimivi proprietate gaudentes — Additamentum I: De curvis elasticis*. Lausanne (1744), reprinted in *Opera Omnia I*, vol. 24, pp. 231–297 (1960).

[8] Géradin, M., Cardona, A.: *Flexible Multibody Dynamics: A Finite Element Approach*, John Wiley & Sons (2001).

[9] Jung, P., Leyendecker, S., Linn, J., Ortiz, M.: A discrete mechanics approach to the Cosserat rod theory - Part 1: static equilibria, *Int. J. Numer. Methods Eng.*, Vol. **85**(1), pp. 31–60 (2011).

[10] Lang, H., Linn, J., Arnold, M.: Multibody dynamics simulation of geometrically exact Cosserat Rods, *Multibody System Dynamics*, Vol. **25**(3), pp. 285–312 (2011).

[11] Lang, H., Arnold, M.: Numerical aspects in the dynamic simulation of geometrically exact rods, *Applied Numerical Mathematics*, Vol. **62**, pp. 1411–1427 (2012).

[12] Linn, J., Stephan, T., Carlsson, J., Bohlin, R.: Fast Simulation of Quasistatic Rod Deformations for VR Applications. L.L. Bonilla, M. Moscoso, G. Platero and J.M. Vega (Eds.): *Progress in Industrial Mathematics at ECMI 2006*, pp. 247–253, Springer (2008).

[13] J. Linn, K. Dreßler. Discrete Cosserat rod models based on the difference geometry of framed curves for interactive simulation of flexible cables. In L. Ghezzi, D. Hömberg and C. Landry (Eds.): *Math for the Digital Factory*, pp. 289–319, Springer, 2017.

[14] J. Linn, T. Hermansson, F. Andersson, F. Schneider. Kinetic aspects of discrete Cosserat rods based on the difference geometry of framed curves. In M. Valasek et al. (Eds.): Proceedings of the *ECCOMAS Thematic Conference on Multibody Dynamics 2017*, pp. 163–176 (2017).

[15] Linn, J., Kleer, M., Schneider, F., Weyh, T. Pena Vina, E.: Messvorrichtung zum Vermessen des Biegeverhaltens einer Probe, Deutsche Patentanmeldung 10 2016 223 900.7, Einreichung: 1.12.2016, Offenlegung: 7.6.2018.

[16] J. Linn: Discrete Cosserat Rod Kinematics Constructed on the Basis of the Difference Geometry of Framed Curves — Part I: Discrete Cosserat Curves on a Staggered Grid. *Journal of Elasticity*, Vol. **139**, pp. 177–236 (2020).

[17] F. Schneider, J. Linn, T. Hermansson, F. Andersson. Cable dynamics and fatigue analysis for digital mock-up in vehicle industry. In: M. Valasek et al. (Eds.): Proceedings of the *ECCOMAS Thematic Conference on Multibody Dynamics 2017*, pp. 763–770 (2017).

[18] F. Schneider, J. Linn: Simulation-based load data analysis for cables and hoses in vehicle assembling and operation.. In: K. Berns et al. (Eds.): Commercial Vehicle Technology 2018. Proceedings, Springer Vieweg, Wiesbaden, pp. 518–529 (2018).

Simulation of an airlay production process

NONWOVEN PRODUCTION PROCESSES

Simulation of fluid-structure-interaction

The application area of nonwoven production, embedded in the field of fluid-structure interactions, offers a multitude of mathematical challenges, since the complexity of the processes renders them unamenable to standard simulation techniques. For numerous key aspects, new models and tools were developed, so that, today, simulation-based solutions for process design and optimization can be generated. The modeling approaches, such as turbulent aerodynamic drag models for fiber dynamics and stochastic surrogate models for nonwoven formation, have opened up interesting research topics for applied mathematics.

NICOLE MARHEINEKE
Trier University

RAIMUND WEGENER
Fraunhofer Institute for Industrial Mathematics ITWM

PARTNERS in the initial BMBF-funded projects ProFil and OPAL:

Academia: FAU Erlangen-Nürnberg [EBERHARD BÄNSCH, GÜNTER LEUGERING],
TU Kaiserslautern [MARTIN GROTHAUS, AXEL KLAR, RENE PINNAU], University Kassel [ANDREAS MEISTER]
Industry: ADVANSA GmbH | Autefa Solutions GmbH | Johns Manville GmbH | IDEAL Automotive GmbH | Oerlikon Neumag

Industrial challenge and motivation

Nonwovens are manufactured in online processes in which the individual process steps are highly coordinated with each other and integrated into a tightly linked chain. There exist a high variety of processes such as spunbond, meltblown or airlay, but they have in common the entanglement and deposition of thousands of filaments or fibers in turbulent airflows leading to the nonwoven irregular cloud-like structure on a conveyer-belt; see Fig. 1 for a spunbond process and Fig. 2 for an airlay process. The application spectrum for nonwovens is extremely broad and ranges from everyday products like diapers and vacuum cleaner bags to high-tech goods like battery separators and medical products. The challenge faced by production is to create a homogeneous fabric quality under chaotic process conditions. The challenge faced by mathematics is to develop models and methods to quantify this quality and to enable process design and optimization. Here, the key technology is the generation of a virtual stochastic microstructure from a process simulation based on first principles.

Figure 1: Spunbond process for production of nonwovens as used for example in baby diapers; spunbond plant of Oerlikon Neumag. Simulation of filament dynamics and airflow in machine geometry: filaments (endless fibers) are visualized in white in front of the airflow where the color indicates the mean velocity magnitude of the turbulent air stream. (Graphic: Simone Gramsch, Fraunhofer ITWM)

Mathematical research

The entanglement and deposition phases of the processes can be viewed as continuum-mechanical, multi-phase problem. However, a closer look on the complexity and variety of scales reveals the hopelessness of a classical monolithic approach and longs, instead, for modeling strategies such as asymptotics and homogenization, along with the generation of surrogate models having a grey box character. In slender body theory solidified filaments and fibers can be described on basis of elastic Cosserat rods exposed to gravity, friction and aerodynamics, [14, 15]. The turbulence enters into the model in an asymptotic limit as white noise, yielding stochastic partial differential equations for the dynamics of the arc-length parametrized time-dependent rod curve $\mathbf{r}$, i.e.,

$$(\varrho A)\, \mathrm{d}\partial_t \mathbf{r} = \{\partial_s(T\partial_s \mathbf{r} - \partial_s((EI)\,\partial_{ss}\mathbf{r}))$$
$$+ (\varrho A)\mathbf{g} + \mathbf{a}(\mathbf{r}, \partial_t\mathbf{r}, \partial_s\mathbf{r}, s, t)\}\mathrm{d}t$$
$$+ \mathbf{A}(\mathbf{r}, \partial_t\mathbf{r}, \partial_s\mathbf{r}, s, t) \cdot \mathrm{d}\mathbf{W}_{s,t} \qquad (1)$$
$$\|\partial_s\mathbf{r}\| = 1.$$

The arc-length constraint enforces local inextensibility and hence the global conservation of length. It turns the scalar-valued inner traction T to an unknown random parameter, i.e., Lagrange multiplier. The system for $(\mathbf{r}, T)$ has a wave-like character due to inertia (line weight (ϱA)) with an elliptic regularization coming from bending stiffness (EI). The aerodynamic effects are covered by a vector-valued space-time Wiener process $\mathbf{W}$ with flow-dependent amplitude $\mathbf{A}$ and a deterministic drag force $\mathbf{a}$. The full simulation of (1) is extremely costly and can therefore only be carried out at an acceptable level of effort for individual filaments/fibers. In order to nonetheless represent the entire microstructure of a nonwoven fabric through a simulation that incorporates the chosen production parameters, stochastic surrogate models are developed that describe the deposition structure of a single filament/fiber on the conveyor belt, instead of the complex dynamic laydown process itself, see e.g., [3],

$$\mathrm{d}\boldsymbol{\eta} = \boldsymbol{\tau}\, \mathrm{d}s \qquad (2)$$
$$\mathrm{d}\boldsymbol{\tau} = (\mathbf{I} - \boldsymbol{\tau} \otimes \boldsymbol{\tau}) \circ (-\nabla V(\boldsymbol{\eta} - \boldsymbol{\gamma})\, \mathrm{d}s + A\, \mathrm{d}\mathbf{W}_s).$$

Figure 2: Airlay process for production of nonwovens as used for example for lightweight components in automotive industry; airlay plant of Autefa Solutions GmbH. Left: Simulation of process with two types of fibers: fibers are visualized in white and yellow in the airflow where the color indicates the mean flow velocity magnitude, machine parts in grey. Right: Virtual nonwoven microstructure (sample column and zoom). (Graphics: Simone Gramsch and André Schmeißer, Fraunhofer ITWM)

In Stratonovich calculus the typical behavior of the arc-length parametrized laydown curve and its tangent $(\boldsymbol{\eta}_s, \boldsymbol{\tau}_s)_{s \in \mathbb{R}_0^+} \in \mathbb{R}^d \times \mathbb{S}^{d-1}$ $(d = 2, 3)$ is modeled by means of a potential V describing the fiber coiling and a production process-dependent reference curve $\boldsymbol{\gamma}$. Fluctuations are taken into considerations by additive noise with amplitude A to the vector-valued Wiener process **W**. The stochastic process $(\boldsymbol{\eta}, \boldsymbol{\tau})$ represents a degenerated diffusion, where the arc-length s takes the role of time in dynamic systems. These laydown models of stochastic differential equations (2) can be very efficiently simulated and allow for the generation of the entire microstructure by superimposing repeated runs. The parameters of the surrogate models are identified from the full simulation of single representative fibers, see, e.g., [6] for spunbond or [4] for airlay.

Dealing with this specific application led to a number of interesting general results in various mathematical fields and inspired ongoing research in different international groups, for an overview and further details we refer to [18]. As examples we point out the following. Existence of solutions and convergence of numerical schemes were proven for semi-discretized versions of the introduced stochastic constrained partial differential equation modeling the rod dynamics in [7, 12]. The surrogate

model for the fiber lay-down stood out by virtue of its significant potential for generalization [1, 10] and analytical results concerning long-term behavior (existence and ergodicity) were achieved, e.g., in [2, 5, 11]. In particular, the aspect of degenerated diffusion increases the challenge of mathematically analyzing this class of models and calls for systematic new developments and extensions, [8, 9].

Implementation

The long-term economic success of the fundamental research could be achieved by application-oriented software development. At Fraunhofer ITWM, the FIDYST Suite was developed consisting of the software tools FIDYST (Fiber Dynamics Simulation Tool) for simulating Cosserat rods in turbulent flows and SURRO (Surrogate Models) for generating complete nonwoven microstructures with build-in parameter identification. FIDYST can process flow data of various types, as well as geometry information in the EnSight format, and therefore works well, for example, in combination with ANSYS Fluent as a CFD tool. In both cases (flow and geometry), interpolations are made between discrete time points for transient information, so that also processes with moveable machine

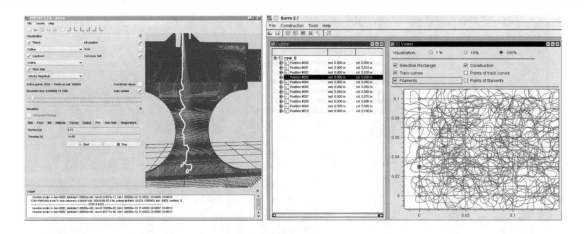

Figure 3: FIDYST Suite. Left: FIDYST GUI with 3D viewer (simulation run). Right: SURRO GUI (deposition with several spinning positions) (Graphics: Simone Gramsch, Fraunhofer ITWM)

parts can be handled. Simulating even large nonwoven structures in only a few seconds, SURRO provides a series of post-processing functionalities for analyzing quality characteristics. For example, fluctuations in weight-per-area can be visualized and quantitatively evaluated using freely selectable scales. Practitioners are frequently interested in the integrated width and length distributions. The same holds true for other quality criteria of the virtual nonwovens, such as strip appraisal. The FIDYST Suite has user-friendly GUIs for initiating simulations, as well as for the accompanying visual simulation guidance and control, cf. Fig. 3 and Fig. 4. It is used for contract research, continuously extended as element of projects in applied basic research, and also licensed to customers.

Nowadays, the FIDYST Suite allows the systematic simulation-based process design for a broad spectrum of nonwoven production processes.

Industrial relevance and summary

Essential impact for the research in this application area of nonwoven manufacturing was achieved with several partners from academia and industry (production and engineering), especially in the projects "Stochastic production processes for the manufacturing of filaments and nonwovens" (ProFil, 2010-2013) and "Optimization of airlay processes"

(OPAL, 2013-2016), both funded by the German Federal Ministry of Education and Research (BMBF) in the initiative "Mathematics for Innovations in Industry". Scientific connectivity was shown in the field of spinning (e.g., asymptotic and numerics of viscous and viscoelastic jets [13, 16, 19], stochastic turbulence modeling [20], kinetic for suspensions [17]). The industrial relevance and the economic benefit of simulation-based solutions for process design and optimization can be seen in the commercial marketing of the Fraunhofer simulation software FIDYST Suite. The license business starts. The eSoftware is still being developed but requires some technical knowledge in use. The previous licensees are therefore rather large international players with corresponding R&D opportunities. The consulting business has been much more pronounced with an order volume of more than EUR 1 million in the past 5 years in projects with an essential use of FIDYST.

References

[1] Borsche, R., Klar, A., Nessler, C., Roth, A., Tse, O.: A retarded mean field approach for interacting fiber structures. SIAM Multiscale Mod. Simul. **15**(3), 1130–1154 (2017)

[2] Doulbeault, J., Klar, A., Mouhot, C., Schmeiser, C.:

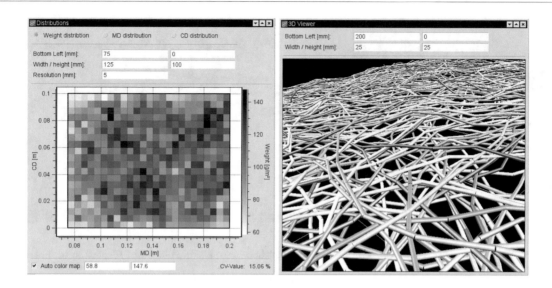

Figure 4: SURRO post-processing. Left: Analysis of weight-per-area distribution. Right: Generation of a 3D microstructure (Graphics: Simone Gramsch, Fraunhofer ITWM)

Exponential rate of convergence to equilibrium for a model describing fiber lay-down processes. AMRX **2013**, 165–175 (2013)

[3] Götz, T., Klar, A., Marheineke, N., Wegener, R.: A stochastic model and associated Fokker-Planck equation for the fiber lay-down process in nonwoven production processes. SIAM J. Appl. Math. **67**(6), 1704–1717 (2007)

[4] Gramsch, S., Klar, A., Leugering, G., Marheineke, N., Nessler, C., Strohmeyer, C., Wegener, R.: Aerodynamic web forming: Process simulation and material properties. J. Math. Ind. **6**(13) (2016)

[5] Grothaus, M., Klar, A.: Ergodicity and rate of convergence for a non-sectorial fiber lay-down process. SIAM J. Math. Anal. **40**(3), 968–983 (2008)

[6] Grothaus, M., Klar, A., Maringer, J., Stilgenbauer, P., Wegener, R.: Application of a three-dimensional fiber lay-down model to non-woven production processes. J. Math. Ind. **4**(4) (2014)

[7] Grothaus, M., Marheineke, N.: On a nonlinear partial differential algebraic system arising in technical textile

industry: Analysis and numerics. IMA J. Num. Anal. **36**(4), 1783–1803 (2016)

[8] Grothaus, M., Stilgenbauer, P.: A hypocoercivity related ergodicity method for singularly distorted non-symmetric diffusions. Integral Equations and Operator Theory (IEOT) **83**(3), 331–379 (2015)

[9] Grothaus, M., Wang, F.-Y.: Weak Poincaré inequalities for convergence rate of degenerate diffusion processes. Annals of Probability **47**(5), 2930–2952 (2019)

[10] Klar, A., Maringer, J., Wegener, R.: A smooth 3d model for fiber lay-down in nonwoven production processes. Kinet. Relat. Models. **5**(1), 57–112 (2012)

[11] Kolb, M., Savov, M., Wübker, A.: (Non-)ergodicity of a degenerate diffusion modeling the fiber lay down process. SIAM J. Math. Anal. **45**(1), 1–13 (2013)

[12] Lindner, F., Marheineke, N., Stroot, H., Vibe, A., Wegener, R.: Stochastic dynamics for inextensible fibers in a spatially semi-discrete setting. Stoch. Dyn. **17**(2), 1750016 (29 pages) (2017)

[13] Marheineke, N., Liljegren-Sailer, B., Lorenz, M., Wegener, R.: Asymptotics and numerics for the upper-convected Maxwell model describing transient curved viscoelastic jets. Math. Mod. Meth. Appl. Sci. **26**(3), 569–600 (2016)

[14] Marheineke, N., Wegener, R.: Fiber dynamics in turbulent flows: General modeling framework. SIAM J. Appl. Math. **66**(5), 1703–1726 (2006)

[15] Marheineke, N., Wegener, R.: Modeling and application of a stochastic drag for fiber dynamics in turbulent flows. Int. J. Multiphase Flow **37**, 136–148 (2011)

[16] Noroozi, S., Alamdari, H., Arne, W., Larson, R.G., Taghavi, S.M.: Regularized string model for nanofibre formation in centrifugal spinning methods. J. Fluid Mech. **822**, 202–234 (2017)

[17] Vibe, A., Marheineke, N.: Modeling of macroscopic stresses in a dilute suspension of small weakly inertial particles. Kinet. Relat. Models. **11**(6), 1443–1474 (2018)

[18] Wegener, R., Marheineke, N., Hietel, D.: Virtuelle Produktion von Filamenten und Vliesstoffen. In: H. Neunzert, D. Prätzel-Wolters (eds.) Mathematik am Fraunhofer Institut, pp. 105–165. Springer, Heidelberg (2015)

[19] Wieland, M., Arne, W., Fessler, R., Marheineke, N., Wegener, R.: An efficient numerical framework for fiber spinning scenarios with evaporation effects in airflows. J. Comput. Phys. **384**, 326–348 (2019)

[20] Wieland, M., Arne, W., Marheineke, N., Wegener, R.: Melt-blowing of viscoelastic jets in turbulent airflows: Stochastic modeling and simulation. Appl. Math. Mod. **76**, 558–577 (2019)

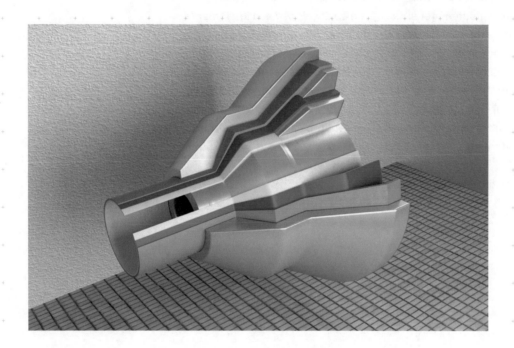

Streamline visualization of particle tracking inside an extruder for polymer processing.

OPTIMIZATION OF MULTIMATERIAL DIES VIA NUMERICAL SIMULATIONS

A multimaterial flowfield simulation for optimizing extruder technologies in polymer processing

Extrusion processes and the corresponding optimization of the extrusion die belongs to the great challenges in polymer processing and is of great importance for the resulting multima- terial extrudates as industrial products, regarding economical aspects as well as their environ- mental impact. The corresponding CFD simulation of the complex flow behaviour poses large challenges to the underlying discretization and solution methods to treat the generalized Navier- Stokes equations which are extended by appropriate nonlinear rheological models. We explain how accurate, robust and highly efficient numerical simulation techniques can be developed and implemented in software modules based on the CFD software FeatFlow which have been realized and applied to real life configuration of industrial relevance as a result of the co-operation between TU Dortmund University (LS3) and IANUS Simulation GmbH.

OTTO MIERKA
MARKUS GEVELER
STEFAN TUREK
TU Dortmund University

PARTNERS

TOBIAS HERKEN, FRANK PLATTE, **IANUS Simulation GmbH, Dortmund**

Industrial challenge and motivation

Co-extrusion of plastic material melts through extrusion dies belongs to one of the standard plastic processing operation units and plays an extremely important role in the overall sequence of production steps since the geometrical construction of the extrusion die is directly responsible for the quality of the final product [1, 2]. Due to the fact that the individual materials may exhibit fundamentally different rheological behaviors [3, 4], their co-extrusion gives rise to complex flow behavior which may only be predicted by CFD simulation tools using modern discretization and solution techniques. Geometrical optimisation of the extrusion die on basis of appropriate CFD simulations allows for the achievement of such a construction design which is able to provide an optimal contactation of the individual material melt streams and thus guarantees the required composition of the final product. Depending on the geometrical complexity of the extrusion dies there are mainly the following two main types which are therefore posing a challenge to potential geometrical optimisation procedures:

- round or flat sheet extruders used for the production of plastic sheets,

- general profile extruders used for the production of shaped products (for example window frames).

According to the first type, the challenges are related to an appropriate geometrical direction of the individual material streams towards the core stream by preventing any kind of (axial or circumferential) unwanted wave formation so to guarantee a uniform thickness of the individual material layers in the final product. Since the different material layers are responsible for the different functions in the product and their respective specific costs might be unproportionally different it is of key importance to guarantee the overall functionality and quality of the product for the possible cheapest production costs by avoiding an unnecessery excess of any of the valuable material streams. The second type of coextrusion features a geometrically more complicated shape of the product consisting of multiple material segments being connected to

each other in a well defined way to provide the designed functionality of the product. Positioning and alignement of the respective co-extrusion inflow streams is the prerequisite for obtaining the required quality of the product. The complexity of the problem is eliminated in the praxis by a sequence of trial and error procedures until a geometrical variant providing the necessary quality of the product is reached. The financial and temporal cost for such a design might be drastically reduced by starting of a more educated guess provided by an appropriate simulation software. Such a simulation software needs to feature the following attributes in order to be helpful in the reduction of the iteration steps:

- has to be robust with respect to the geometrical variations

- has to be able to take into consideration the typical shear and temperature dependent rheology of the underlying materials

- has to be able to represent the dynamic interface of the material streams

- has to be efficient, what means to be fast enough to provide the necessary results in reasonable time windows with respect to a reasonable precision of the obtained results.

With the aims of fulfilling the above listed conditions there has been developed an extension Extrud3D to the open-source software package FeatFlow (*www.featflow.de* and see [5]) which has been integrated into the industrial software platform *StrömungsRaum* developed by the IANUS Simulation GmbH. IANUS Simulation GmbH, a Dortmund-located SME, which is a cooperation partner of TU Dortmund University since many years is bringing simulation technology to small, medium and large industrial companies for over a decade. Customers' applications cover chemical engineering, including aspects from Computational Solid Mechanics and particularly Computational Fluid Dynamics. With their system solutions, IANUS Simulation GmbH fuses access to high-end simulation hardware, software and expertise. With their 'Simulation-as-a-Service' cloud computing platform *StrömungsRaum*, they provide individual web applications for their customers covering all aspects of processing the simulation including CAD construction, defining boundary conditions, the automatic extraction of the fluid space, geometry processing and discretisation, deployment of the simulation to a suitable data-centre, automatic analysis and processing of results. All these steps can be performed using only a webbrowser-capable device over the internet while continuous improvement of the underlying simulation tasks is based on frequent meetings between the IANUS Simulation GmbH and TU Dortmund teams.

Mathematical research

Numerical flow simulation of shear- and/or temperature dependent materials is a mathematical challenge by itself and extended into a framework of multimaterial flows imposes really hard expectations on the used solver. Due to the fact that the materials are shear rate (i.e. velocity-gradient) dependent [3, 4], numerical frameworks based on higher order FEM discretization are in advantage [6, 7, 8] and guarantee a higher order of convergence than traditional finite volume or finite difference solvers. This is also the reason why the open-source FeatFlow software has been chosen and extended with the additional ingredients. Based on the two main types of applications mainly differing in the geometrical complexity there have been developed two types of simulation variants being based on the first hand on ALE (Arbitrary-Lagrangian-Eulerian) methods allowing a highly accurate mesh-aligned interface tracking of the present phases - since the topology of the materials is preserved - and on the second hand on a robust and stable particle-tracking based numerical method. A brief description of both of these methods and their respective extension is as follows.

According to the first framework a resolved interface tracking numerical simulation tool is realized according to which the computational mesh is dynamically moving (ALE) so to follow the actual position of the interface which due to the used higher order finite elements (triquadratic Q_2) allows for a highly accurate approximation of the interface curvature even on a single element having one face

aligned with the interface (detailed description is available in [9]). Additional benefit inherited from the interface aligned meshes is the attribution of different material properties on the level of elements which practically has the potential to approximate each material stream by only a very few number of coarse grid elements and guarantees then a very fast convergence of the related geometric multigrid solver. In this respect there is also no additional restriction due to the approximation of the local shear rate (and the resulting local viscosity) and their respective spatial convergence which is due to the employed higher order finite elements. A typical simulation setup visualizing a slice of the estimated final material stream positions with the respective computational mesh of a 5 layer multimaterial flow is displayed in Fig.1.

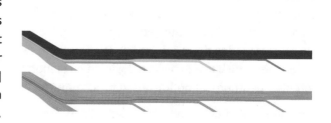

Figure 1: Typical simulation of a multimaterial flow with a moving mesh framework. Result of the final steady state material distribution (top) and the respective computational mesh (bottom).

According to the second framework the simulations transform to an iterative solution scheme conerging on the basis of the altered updates of flow field and material composition. In particular, for a guessed material composition the flow field is being updated by attributing elementwisely constant material properties, and on the obtained flow field a special particle tracking simulation is performed in such a way that from each center of elements of the flow mesh a backtracing of the corresponding streamline is performed in order to identify the related inflow stream with which it is then associated in the next flow field computation iteration. A demonstration of the solution strategy is demonstrated in Fig.2. Clear benefit of this framework is related to its robustness, because it supports the use of general

Figure 2: Typical simulation of a multimaterial flow with a particle tracking framework. Result of the final steady state material distributions (left) with the backward tracing of streamlines and a resulting nonsymmetric velocity distribution due to the different rheological parameters of the involved materials.

meshes without the necessity of being aligned with the interfaces and in general a convergence of the scheme is obtained after 3 to 5 iteration steps.

Implementation

According to the previously described mathematical descriptions, both numerical frameworks have been implemented up to a certain degree of automation into the production software Extrud3D as an extension for multimaterial flows. Since the first (moving mesh) framework requires the construction of interface aligned meshes, here the computational meshes are created in a semi-automated way, meaning that a manually designed 2D blockmesh is extruded into full 3D (in a perpendicular way for flat sheet extrudes, or rotationally for round extruders). Due to the offered flexibility of the second framework the achieved extent of automation is much higher. In particular, the extension has been embedded into the standard automation mechanism of Extrud3D, so that an automatic coarse-box mesh is created, which is exposed to a few mesh deformation steps after which only the mesh cells are kept which are covering the fluid geometry. The solution of the momentum equation, as well as the particle tracking, is performed on basis of the such obtained coarse mesh refined to the necessary resolution level and by the well established Fictitious Boundary Method (FBM) [10] which filters out the 'solid'

degrees of freedom and in the context of the particle tracking restricts the streamlines only to the fluid domain. Both simulation frameworks have been gradually integrated into the *StrömungsRaum* platform, which in this particular case was associated with the challenges to implement the automation of preprocessing, solving and postprocessing steps. The *StrömungsRaum* offers the users of the platform to upload CAD descriptions (see Fig.3) of various die configurations, specify the inflow and outflow surface segments, prescribe the corresponding flow rates and to specify the underlying material properties from the available material database linked to the web application. At this stage the simulation case can be submitted for simulation and upon obtaining the simulation results the postprocessing software Reporter (developed by IANUS Simulation GmbH) extracts and compiles the user defined postprocessing items into a simulation report for the users of the platform.

Industrial relevance and summary

The fundaments of the numerical techniques have been developed at the TU Dortmund University as part of the DFG research projects SPP 1740:TU 102/53-1 and KU 1530/13-11 and have been integrated into the industrially used software package Extrud3D developed by IANUS Simulation GmbH. By the help of the corresponding

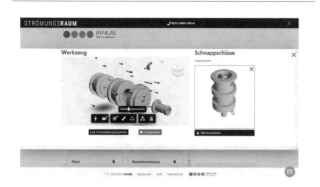

Figure 3: Typical visualization of die geometries in the *StrömungsRaum* platform of IANUS Simulation GmbH.

extensions not only the simulation of existing extrusion dies has become possible, but even the possibility of optimization of individual geometrical realizations has been reached. Due to the developed simulation tool novel construction possibilities have become accessible for the producers of multimaterial extrusion dies according to which a quantitatively higher quality of products can be achieved by guaranteeing the functional properties of the multimaterial extrudates brought to the market. The other industrially significant effect of the developed approach is that the die-construction process may be dramatically accelerated by skipping the traditional experimental adjustment phase (trial/error) which allows faster construction delivery times and at the same time also the amount of test materials wasted for these purposes may be drastically reduced, which besides of economical aspects relaxes the environmental impacts, too.

Acknowledgement

The authors acknowledge the support by the Federal Ministry of Education and Research (BMBF project no. 05M13RDC, ExtremSimOpt, and project no. 05M16PEA, BlutSimOpt).

References

[1] Ch. Hopmann and W. Michaeli, *Extrusion Dies for Plastics and Rubber*, Hanser, 2017, 978-1-56990-623-1.

[2] S. Kainth, *Die Design for Extrusion of Plastic Tubes and Pipes*, Hanser, 2018, ISBN: 978-1-56990-672-9.

[3] K.R. Rajagopal, *Mechanics of non-Newtonian fluids, Recent Developments in Theoretical Fluid mechanics*, Pitman Research Notes in Mathematics 291, eds. G.P. Galdi und J. Necas, 129-162, Longman, 1993.

[4] M. T. Shaw, *Introduction to Polymer Rheology*, Wiley, 2011, ISBN: 978-0-47038-844-0

[5] S. Turek, *Efficient Solvers for Incompressible Flow Problems: An Algorithmic and Computational Approach*, LNCSE 6, Springer, 1999.

[6] F. Brezzi and M. Fortin, *Mixed and Hybrid Finite Element methods*, Springer, Berlin, 1986, ISBN: 978-1-4612-7824-5.

[7] V. Girault and P. A. Raviart. *Finite Element Methods for Navier-Stokes equations*, Springer, 1986. Berlin-Heidelberg, ISBN: 978-3-642-64888-5.

[8] P. M. Gresho, M and R. L. Sani, *Incompressible Flow and the Finite Element Method: Advection-Diffusion and Isothermal Laminar Flow*, Wiley, 1998. ISBN 0 471 96789 0

[9] S. Turek, O. Mierka, Numerical Simulation and Benchmarking of Drops and Bubbles, book chapter in *Handbook of Numerical Analysis*, 2019, Elsevier, ISSN 1570-8659, doi:10.1016/bs.hna.2019.09.001.

[10] R. Münster, O. Mierka, S. Turek: Finite element-fictitious boundary methods (FEM-FBM) for particulate flow, *Int. J. Numer. Meth. Fluids*, **69**, pp:294-313, 2012.

Energy optimization of wine fermentation

ENERGY CONSERVATION FOR WINE FERMENTATION

Nonlinear model predictive control for placement and real-time control of cooling plates

A large amount of energy is used in the production of wine, which has to be cooled during fermentation in order to produce a multiplicity of aromas. If not cooled, the added yeast transforms the sugar content of the raw material, the must, into alcohol without much of a taste, in very short time. Thus the cooling during the fermentation process is a major component in the production process of wine. Traditionally, cooling is performed by usage of an electrical refrigerator according to a rule-of-thump-cooling scheme. In this collaborative project, we employ methods from modeling, simulation and optimization, in order to determine optimal cooling profiles, which are able to reduce the energy consumption by a factor of two without sacrificing the taste, which is the main reason to consume wine. Thereby, the main objective is to make the process more energy efficient.

The reduction of the cooling costs can be realized by improving the cooling plate placement and size and by real-time temperature control using an economic nonlinear model predictive control strategy.

The resulting mathematical methods and algorithms are based on theoretical research in optimal control of differential equations and its generalization to integro-differential equations as well as in shape optimization. These topics are at the core of the research activities of the research group of Volker Schulz, Trier university. This research has been performed in collaboration with the universities of Würzburg and Geisenheim and Dienstleistungszentrum Ländlicher Raum (DLR) Mosel, Bayerische Landesanstalt für Weinbau und Gartenbau (LWG), fp sensor systems GmbH.

JAN BARTSCH
University of Würzburg, Institute for Mathematics

ALFIO BORZI
University of Würzburg, Institute for Mathematics

JONAS MÜLLER
Geisenheim University, Modeling and Systems Analysis

CHRISTINA SCHENK
Trier University, Department of Mathematics,

DOMINIK SCHMIDT
Geisenheim University, Modeling and Systems Analysis

VOLKER SCHULZ
Trier University, Department of Mathematics

KAI VELTEN
Geisenheim University, Modeling and Systems Analysis

PARTNERS

ACHIM ROSCH, Dienstleistungszentrum Ländlicher Raum (DLR) Mosel, Department of Winegrowing and Enology
MICHAEL ZÄNGLEIN, Bayerische Landesanstalt für Weinbau und Gartenbau (LWG), Department of Winegrowing
PETER FÜRST, fp sensor systems GmbH

Industrial challenge and motivation

The profit of an industrial company can be increased by reducing the production costs but maintaining the quality at the same time. Therefore, the application of mathematical modeling, simulation and optimization techniques establishes more and more in industry. In the context of fermentation processes, this is the main objective of the project RŒNOBIO. There is a high potential for saving energy in the process of making wine. In 2009, the energy consumption generated 0.08% of the global greenhouse gas emissions or in other words about 2 kg/0.75 l bottle [1]. For instance in California the annual energy requirements of the wine industry are located at 400GWh. This makes it the second highest energy consumer in the food industry [2]. Thereby, the control of the fermentation temperature plays a crucial role [3]. Therefore, the minimization of the energy needed for cooling during wine fermentation matters.

Figure 1: A typical wine tank .....

Sugar being transformed into alcohol is an exothermic reaction. This means that the produced heat has to be dissipated as the temperature development is very important for the yeast. If the fermentation temperature is too high, yeast cells will die. However, in the phase where oxygen is available, even more heat is generated. Moreover, the change of temperature due to the tank's environment and the cooling element which is switched on or off based on the control input are taken into account. Thus, a novel complete model including all relevant processes and species had to been developed, which also describes the growth of yeast as an integro-differential process and considers the dying phase due to alcoholic poisoning, as well. Furthermore, a method for adapting model parameters to measurements had to be developed as well as an optimal control strategy. A particular challenge is due to the scarcity of available data. Thus, an economic nonlinear model predictive control (ENMPC) strategy, minimizing the cooling energy, needed during the fermentation process by controlling the fermentation temperature and maintaining the wine quality (especially of white wine), with the performance of parameter and state estimation (PSE) is solved. This means that whenever new measurements are available the model parameters and states are estimated again and the new control input is computed.

Mathematical research

Reducing the energy consumption during wine fermentation, it is very important not to sacrifice quality. In this regard, we developed a real-time optimization strategy regarding these two objectives and controlling the temperature using a switching structure formulation. It combines an economic nonlinear model predictive control approach with the consistent estimation of states and parameters. As a basis, other investigations had to be made, namely regarding parameter identification, function identification for the death term of the yeast cells related to the accumulated ethanol concentration, 3D-multiphysics-simulation and optimization of the cooling plate size and placement. The 3D-simulation showed that in the course of active fermentation it is sufficient to use a 0D-model, while natural convection driven cooling, e.g. during storage or pre-fermentation, necessitates 3D-simulations (Fig. 2). The real-time optimization strategy was validated for the derived realistic model in the context of several experiments, run in collaboration with our public research and industry partners (Fig. 3). our objective in the optimization problem, formulated in the following, consists in the minimization of the energy, needed for cooling by controlling the fermentation temperature, in combination with the maintenance of the quality and constraints on the maximum temperature, the control input ω and the residual sugar, expressed by a boundary condition. The considered optimization problem is of the form

$$\min_{x,\omega} \quad \gamma_1 \int_{t_c}^{t_c+\Delta t} \omega(t)dt + \gamma_2 \int_{t_c}^{t_c+\Delta t} (S(t) - \hat{S}(t))^2 dt$$

$$\text{s.t.} \quad \frac{dx(t)}{dt} = f(t, x(t), p) \quad \text{(fermentation dynamics)}$$

$$\frac{dT(t)}{dt} = \alpha_1 \frac{dE(t)}{dt} - \alpha_2 \frac{dO_2(t)}{dt}$$

$$- \alpha_3(T(t) - T_s)\omega(t) - \alpha_4(T - T_{\text{ext}})$$

Here Δt denotes the planning time horizon starting from the current time t_c. The objective is a weighted average between energy consumption and the goal of obtaining a sugar time profile ($S(t)$) close to a desired profile. The constraint consists on the one hand of the fermentation process dynamics and on the other hand of the temperature dynamics influenced by the anaerobic production of ethanol (E), the consumption of oxygen (O_2) by the yeast in the beginning of the process, the cooling action by partly (ω) opening a valve for cooling soil (T_s) and the temperature exchange with the cellar. The biologically appropriate fermentation dynamics includes the growth of the yeast cells and split up of mother cells to smaller mother cells and daughter cells, which amounts to an integro-differential component in the mathematical model for this process. The remaining species (ethanol, carbon dioxide, oxygen, sugar, nitrate) are coupled with the yeast process in a Michaelis-Menten-type of dynamics. Furthermore, a model for the yeast dying process has been developed for the first time. The mathematical challenge arose from the specific mathematical modeling of the process and the handling of the nonlinear integro-differential model within an optimization framework including appropriate and mathematically founded approximations. More details on the results can be found in the publications [4, 5].

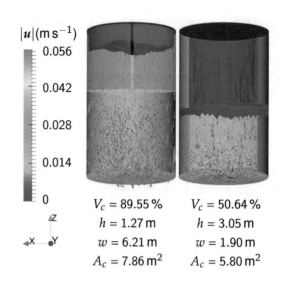

$|u|\,(\mathrm{m\,s^{-1}})$

0.056

0.042

0.028

0.014

0

	$V_c = 89.55\%$	$V_c = 50.64\%$
	$h = 1.27\,\mathrm{m}$	$h = 3.05\,\mathrm{m}$
	$w = 6.21\,\mathrm{m}$	$w = 1.90\,\mathrm{m}$
	$A_c = 7.86\,\mathrm{m^2}$	$A_c = 5.80\,\mathrm{m^2}$

Figure 2: Differences in cooling performance during storage of wine depending on cooling jacket dimensions (light-blue) represented by the iso-surface (blue) of the cooled volume (V_c with $\Delta T = 1\mathrm{e}{-}3\,\mathrm{K}$) after 30 min of cooling (left: low and wide, right: high and thin).

Industrial relevance and summary

This research work was part of the project RŒNOBIO (Robust energy-optimization of fermentation processes for the production of biogas and wine), funded by the German Federal Ministry of Education and Research with contract number 05M2013UTA running from 2013 to 2017. It was a collaborative project with devoted researchers from Geisenheim, Trier and Würzburg and additional partners consisting of public research and industry partners. This project required a close interaction in between industry/public research and academia. As this succeeded, it led to fruitful discussions, assured successful results and initialized promising ideas for further research projects regarding food processes. It showed the dramatic potential for energy savings in wine fermentation. The implementation for usage also in small wineyards is planned in the form of a planning app on smartphones. The interaction between the wine experts and the researchers led to successful results, i.e. tasty wines with less energy needed for cooling during fermentation.

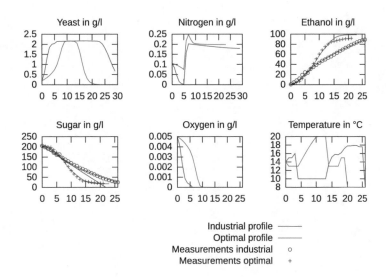

Figure 3: Results of a Riesling experiment: Comparison of trajectory with industrial controller (purple) and trajectory with MPC-controller (blue)

References

[1] Smyth, M., et al., 2011. Solar Energy in the Winemaking Industry. Green Energy and Technology, Springer.

[2] Galitzky, C., et al., 2005. Benchmarking and self-assessment in the wine industry. In: Proceedings of the 2005 ACEEE Summer Study on Energy Efficiency in Industry.

[3] Freund, M., 2008. Energy and water saving in wine processing. Obst- und Weinbau 144 (19), 4–7.

[4] Christina Schenk, Volker Schulz, Achim Rosch and Christian von Wallbrunn, 2017. Less cooling energy in wine fermentation – A case study in mathematical modeling, simulation and optimization,Food and Bioproducts Processing vol. 103, 131–138.

[5] Christina Schenk 2018. Modeling, Simulation and Optimization of Wine Fermentation. Dissertation, Trier University.

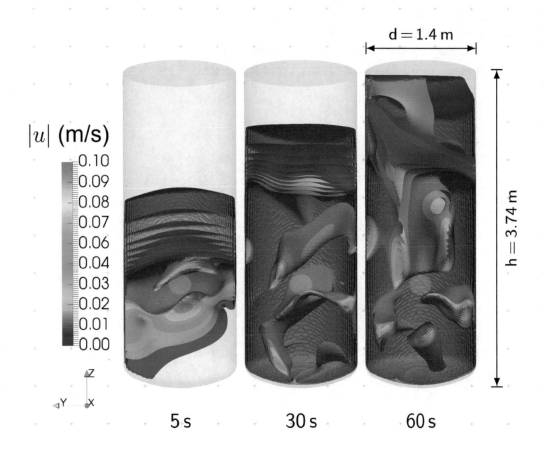

Simulation of fast-rotating fluids in a wine tank

© Springer Nature Switzerland AG 2021
H. G. Bock et al. (eds.), *German Success Stories in Industrial Mathematics*,
Mathematics in Industry 35, https://doi.org/10.1007/978-3-030-81455-7_13

A CFD approach for fast-rotating machinery

The beverage and wine industries routinely conduct blending of various liquids or homogenization of additives in large vessels equipped with fast-rotating propeller mixers. Mixing times, i.e. propeller run times, are usually based on empirical experiences. Mathematical modeling and simulation could be used to predict mixing times in such situations, but common methods either provide only approximate solutions or are computational unfeasible. Hence, this project addresses the challenge to allow accurate long time simulations of fast rotating propeller mixers in large vessels. The developed method can be used to study transient mixing processes in large tanks by fast-rotating propeller mixers.

JONAS MÜLLER
DOMINIK SCHMIDT
KAI VELTEN
Geisenheim University, Modeling and Systems Analysis

PARTNERS

MARCEL SZOPA, **Henkell & Co. Sektkellerei KG, Wiesbaden**

Industrial challenge and motivation

Fast-rotating mixers are used in the wine and beverage industries to homogenize large volumes of liquids. Typical mixers are movable and can be used in tanks of various sizes [2]. However, even the largest producers, such as our partner the Henkell & Co. Sektkellerei KG, rely on empirical practices with regard to mixing times with substantial consequences for the energy consumption in the industry. The question posed by industry partners within the framework of a larger project regarding energy optimization in the wine and biogas industry, is how long it takes to sufficiently mix a given tank with a given mixer. The Department of Modeling and Systems Analysis at Hochschule Geisenheim University is specializing in applied computational fluid dynamics (CFD) modeling in the food and beverage industry with a strong emphasis on research topics relevant to industrial-scale wineries. The two traditional approaches in CFD for rotating machinery are the moving reference frame method (MRF) [3] and the sliding mesh method (SM) [1].

While the former is only appropriate for steady-state simulations, the latter is prohibitive in its computational cost when mixing times exceed a few seconds. A robust method using CFD models to predict mixing times in large tanks became apparent and was the initial motivation to conduct the subsequently described research.

Mathematical research

Simulating small fast rotating propellers in large wine tanks leads to a large 'dimensional gap' in a CFD domain. To accurately resolve the propeller geometry, small grid cells are necessary in a region where the highest fluid velocities are expected. In CFD simulations the discretization error can be limited by keeping the Courant-Number below one. In practice, this can be described as keeping the time-step low enough, that a fluid particle passes no more than one grid cell during one iteration. As a consequence, high fluid velocities in regions with small cell sizes determine the computational cost of the whole simulation.

To address this issue we propose to simplify the problematic propeller region within the computational domain and replace it with a boundary condition representing the impulses introduced by a propeller mixer (Figure 1). We assume the delivered impulse of a fast rotating propeller to be periodic and depending on the current propeller orientation. Aiming for the extraction of a simple geometric shape from the computational domain, we further assume this periodicity to stabilize in the near field of the propeller.

In a preliminary study of the propeller zone, both current CFD approaches are combined by first initializing a steady-state flow field using an MRF simulation followed by a time-dependent SM

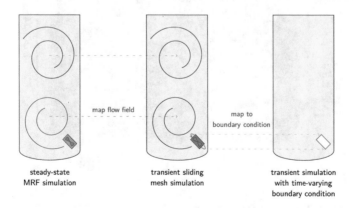

steady-state
MRF simulation

transient sliding
mesh simulation

transient simulation
with time-varying
boundary condition

map flow field

map to
boundary condition

Figure 1: Workflow of the proposed approach to replace the propeller by a time-varying boundary condition with data derived from previous transient sliding mesh simulation, that starts from a flow field initialized by a steady-state MRF simulation.

simulation for enough propeller rotations to establish periodic stability. Cell data are then sampled over one full rotation from the surface of a cylindrical field in the near distance to the propeller, that is already available within the domain, as it is used as the reference frame in the MRF method and as the sliding mesh surface in the SM simulation. To simulate the mixing process from a non-agitated state, a transient simulation is set up where the mesh region within the cylindrical reference frame is removed and the sampled data is mapped to the new surface using a time-varying boundary condition. Hence, the region with the highest velocities and the smallest cell sizes is completely removed from the domain and replaced with a much less computational intensive transient boundary condition. This transient boundary condition approach allows up to 20-fold speed-ups when compared to transient simulation times of the sliding mesh. More details about the developed methodology can be found in Müller et al. (2018) [4].

Implementation

Based on several meetings with our partner we identified the need for a simple smaller scale experiment to develop a methodology for addressing the large scale mixing problem. Validation experiments for the new CFD method were conducted at Hochschule Geisenheim University. A fast-rotating movable mixer, similar to the ones used by our industrial partner, was installed in a commercial 6000 L wine tank filled with water up to 0.5 m below the tank lid and a HD video camera was installed to observe the water surface. Repeated time measurements quantified the arrival of the mixer impulse on the liquid surface. Tank and mixer were modeled according to the manufacturers technical drawings using CAD software. During the setup of the CFD simulation rheological properties of the liquid, as well as operational properties of the mixer were implemented. Subsequently, preliminary CFD simulations were conducted as described above using OpenFOAM [6] to extract the periodic data in the near field of the propeller. In consultation with

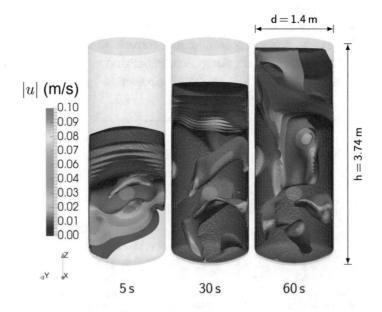

Figure 2: Results of the pilot study showing the impulse propagation induced by the transient boundary condition over time, indicated by iso-surfaces of velocity magnitudes within the tank. Simulations results were used to validate the time point of surface wave appearance. The hollow cylinder inside the tank are the inner boundary patches resulting from the removal of the internal cells needed in the pre-processing to resolve the propeller geometry.

the industrial partner we decided on using the free open-source CFD toolbox OpenFOAM [6] for this prototypical project. Comparing the experimental data and the time-dependent propagation of the flow field simulated using the boundary condition approach, a good accordance could be established. Notably, the sudden arrival of a visibly moving liquid zone (defined as $>0.1\,\mathrm{m\,s^{-1}}$) on the surface concurred after approximately 60 seconds in both, experiment and CFD simulation (Figure 2).

Industrial relevance and summary

Progress has been made in answering the specific problem regarding mixing in large tanks arising from the industry. A new CFD method was developed using a mapped, time-varying boundary condition in place of the propeller mixer, significantly reducing the computational cost [4]. The newly developed approach allows for the first time to analyze specific tank-mixer configurations over time frames relevant to the industry using CFD. Potential industrial applications include the optimization of mixer placement, the dimensioning and design of new mixers for different size tanks and, by combining the method with a model for mixing quality, the calculation of minimum mixing times. Strong connections to industry partners and collaborating researchers from the RŒNOBIO project (Robust energy-optimization of fermentation processes for the production of biogas and wine), funded by the German Federal Ministry of Education and Research, led to a continuous and constructive dialog of people with very diverse mathematical and engineering backgrounds. This exchange helped to focus the research involving two Ph.D. students (J. Müller and D. Schmidt) at Hochschule Geisenheim University. Their results showed, that the existing CFD methods for rotating objects were not appropriate for the specific problem at hand. The institute developed and validated a new method modeling the mixing in large tanks and perpetually shared the results with their academic and industrial partners. These results were presented at a CFD conference in Portugal [5] and at different venues in Germany, garnering interest from CFD engineers and technical directors of large wineries. The RŒNOBIO project and

contacts from the workgroup to industry partners brought people of very different backgrounds together and unfolded synergetic effects that allowed for this project to emerge. Building on the knowledge of the group, it was possible to develop a method that finds practical use in the industry, adds another tool to the CFD engineer's toolbox, and may result in significant energy savings when applied in large wineries or related fields.

Acknowledgements

The authors acknowledge the support by the Federal Ministry of Education and Research (BMBF project no. 05M13RNA, Robust energy optimization of fermentation processes for the production of biogas and wine (ROENOBIO)).

References

[1] A. Bakker, R. D. LaRoche, M.-H. Wang, and R. V. Calabrese. Sliding mesh simulation of laminar flow in stirred reactors. *Chemical Engineering Research and Design*, 75(1):42–44, 1997.

[2] R. B. Boulton, V. L. Singleton, L. F. Bisson, and R. E. Kunkee. *Principles and practices of winemaking*. Springer Science & Business Media, 2013.

[3] W. Bujalski, Z. Jaworski, and A. Nienow. Cfd study of homogenization with dual rushton turbines—comparison with experimental results: Part ii: The multiple reference frame. *Chemical Engineering Research and Design*, 80(1):97–104, 2002.

[4] J. Mueller, D. Schmidt, and K. Velten. Transient boundary condition approach for simulating mechanical mixing in large wine tanks. *Computers and Electronics in Agriculture*, 150:143–151, July 2018.

[5] J. Mueller and K. Velten. Simulating tank-mixer problems using a time-varying mapped fixed value approach. In *11th OpenFOAM Workshop, Guimaraes, Portugal*, 2016.

[6] H. G. Weller, G. Tabor, H. Jasak, and C. Fureby. A tensorial approach to computational continuum mechanics using object-oriented techniques. *Computers in physics*, 12(6):620–631, 1998.

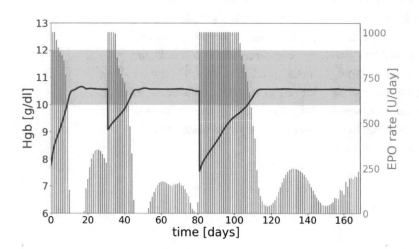

Optimal EPO treatment for a patient with sudden and unforeseen bleedings

PERSONALIZED MEDICINE – OPTIMIZED EPO DOSING

A nonlinear model predictive control method for individualized EPO doses

Currently, epoetin alfa (EPO) doses for hemodialysis patients are usually prescribed using dosing protocols, where doses can be looked up based on previous hemoglobin (Hgb) measurements. The aim is to stabilize Hgb levels within the narrow target window of 10 – 12 g/dl. This is a delicate objective, as patients' response to treatment varies highly and doses should be as small as possible to mitigate drug-related hazards. The Renal Research Institute (RRI) together with the University of Graz developed a predictive model of erythropoiesis which takes EPO administration into account. This model possesses a number of personalized parameters and parameter estimation is done for individual patients at the institute. A joint project was initiated to develop and implement a nonlinear model predictive control method for computation of individualized EPO doses using the previously developed erythropoiesis model.

FRANZ KAPPEL
Institute for Mathematics and Scientific Computing, Karl-Franzens University of Graz

STEFAN VOLKWEIN
Department for Mathematics and Statistics, University of Konstanz

PARTNERS

DORIS FÜRTINGER, Fresenius Medical Care Deutschland GmbH, Bad Homburg
SABRINA ROGG, Fresenius Medical Care Deutschland GmbH, Bad Homburg
PETER KOTANKO, Renal Research Institute New York, New York, USA, Icahn School of Medicine at Mount Sinai, New York, USA

Personalized Medicine – Optimized EPO Dosing

Currently, epoetin alfa (EPO) doses are often prescribed using empirical algorithms which are in general charts, where doses can be looked up based on a current hemoglobin measurement. The aim is to stabilize hemoglobin levels within a narrow target window. This is a delicate objective, as patients' response to treatment varies highly and doses should be as small as possible to mitigate drug-related hazards. The Renal Research Institute (RRI) developed a predictive model of erythropoiesis which takes EPO administration into account. This model possesses a number of personalized parameters and parameter estimation is done at the institute individually for each patient. A joint project was initiated to develop and implement a nonlinear model predictive control method for computation of the EPO doses based on the previously developed erythropoiesis model.

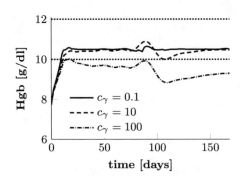

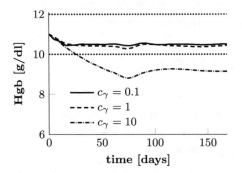

Figure 1: Optimal Hgb curves for different values for c_γ for patient 1 (upper plot) and for patient 2 (lower plot). The dotted lines mark the desired target range; see [8, Figure 4].

Implementation of the initiative

The collaboration between the University of Konstanz and the RRI started already in 2014 with several master theses [1, 2, 6, 7]. A research associate in the working group of Prof. Volkwein developed and implemented the control algorithm while the institute provided anonymized model parameters of several patients to test the algorithm with. The research associate visited the institute twice and regular video meetings were held between the project partners allowing a permanent dialogue.

The optimization approach

The presented results are mainly based on our recently published work [8]. The goal was to develop a controller scheme that is fully personalizable. The used model of erythropoiesis consists of five hyperbolic partial differential equations for the different cell stages a cell passes through when becoming an erythrocyte; cf. [3, 4]. Time delay in the sense that the cells remain about two weeks in the bone marrow before getting released into the bloodstream makes therapy difficult. The goal is to stabilize the hemoglobin (Hgb) around a desired value in order to bring and keep the Hgb into a given target window. Values in that window are considered to be safe. The control input is given by the drug administration for EPO that enters into the coefficients of the hyperbolic equations in a highly nonlinear manner. The optimal control problem is formulated for an assumed continuous but piecewise constant EPO administration. This means that the optimal administration rate over time is searched. The optimal control problem is formulated for the number of RBCs and not for the Hgb. Especially, for each patient, prior to the optimization a desired total amount of RBCs is computed. The structure of the control process is illustrated in Figure 2; cf. [8].

If the erythrocytes population density is known, the total RRC population can be computed by

$$P(t) = \int_{x_a}^{x_b} y_5(t, x)\, \mathrm{d}x \quad \text{for } t \in [0, T].$$

The objective is of tracking type, where we penalize the difference between the total RBC population from

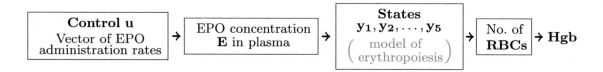

Figure 2: Schematics of the control structure; cf. [8, Figure 1].

the desired population P^d over a time interval of interest

$$\hat{J}(u) = \frac{\sigma}{2} \int_0^T \left| P(t) - P^d \right|^2 dt + \frac{\gamma}{2} \sum_{j=1}^m |u_j|^2,$$

where the total RBC population $P(t)$ depends on the erythrocytes population density y_5 and σ, γ are given positive cost weights. Summarizing we get a nonlinear and therefore nonconvex optimal control problem. However, this problem is not considered as an open-loop problem since unforeseen events and disturbances can occur. In practise, predicted and measured Hgb values differ. This has to be taken into account by utilizing a closed-loop controller. Furthermore, patient parameters are usually not constant during the treatment. Thus, modifications of the model have to be included, too. In our approach we therefore apply a (nonlinear) model predictive control (NMPC) method (see, e.g., [5]) to get the optimal EPO drug administration in a closed-loop manner. In each step of the NMPC algorithm an open-loop optimal control problem is solved on a small horizon. Here we utilize a projected quasi-Newton method so that we avoid the computation of second derivatives while still ensuring fast local convergence properties.

Results and achievements

We successfully found an optimization setting which works for many patient data sets. Here, the developed NMPC algorithm has been able to control the patients' number of erythrocytes very satisfactorily. In Figure 1 the total RBC population is presented over time for two different patients, where the control cost weight γ is given as

$$\gamma = \frac{c_\gamma}{0.28}.$$

For the choice of the parameter σ we refer to [8]. For both patients the choice $c_\gamma = 0.1$ (i.e., $\gamma \approx 0.36$) is the right choice for control costs penalization in the sense that it allows to accurately stabilize the Hgb around 10.5 g/dl. The corresponding EPO rates can be seen in Figures 4 and 5.

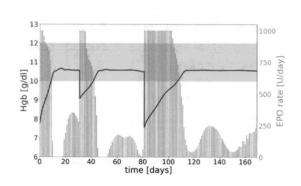

Figure 3: Optimal EPO treatment for patient 1 subject to sudden and unforeseen bleedings at day 30 (7.5 g/dl) and 80 (8 g/dl).

The numerical tests have shown that the algorithm can handle the mentioned time delay as well as simulated bleedings. This can be seen in Figure 3.

On-going work

On-going work is devoted to the restriction that EPO can only be administered during dialysis treatments and it remains to test the algorithm for all available data sets. Moreover, we are going to study the nonlinear system dynamics in a more detail with respect to controllability and observability in order to improve the performance of the method even more. More precisely, we will apply extended dynamic mode decomposition (eDMD) to derive specific linearizations for the five coupled hyperbolic

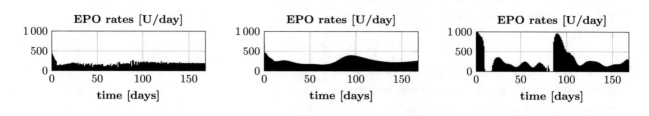

Figure 4: Optimal EPO rates for different values for c_γ for patient 1; see [8, Figure 5].

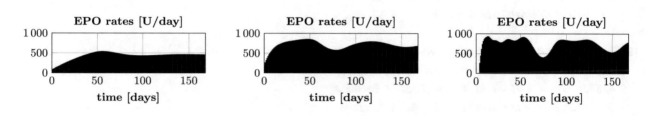

Figure 5: Optimal EPO rates for different values for c_γ for patient 2; see [8, Figure 6].

partial differential equations. Preliminary results are recently achieved for a simplified model in [9].

Acknowledgement

S. Rogg and S. Volkwein acknowledge partial support by the Renal Research Institute, New York, USA.

Partners

» Doris Fuertinger, Fresenius Medical Care Deutschland GmbH, Bad Homburg, Germany

» Franz Kappel, Renal Research Institute, New York, USA

» Peter Kotanko, Renal Research Institute, New York, USA

» Sabrina Rogg, Fresenius Medical Care Deutschland GmbH, Bad Homburg, Germany

» Stefan Volkwein, University of Konstanz, Konstanz, Germany

References

[1] D. Beermann. *Reduced-Order Methods for a Parametrized Model for Erythropoiesis Involving Structured Population Equations with One Structural Variable*. Diploma thesis, University of Konstanz, 2015. See http://nbn-resolving.de/urn:nbn:de:bsz:352-0-283508.

[2] F. Binder. *Sensitivity analysis for a parametrized model for erythropoiesis involving structured population equations with one structural variable*. Master thesis, University of Konstanz, 2016. See http://nbn-resolving.de/urn:nbn:de:bsz:352-0-361615.

[3] D. Fuertinger. *A model of erythropoiesis*. PhD thesis, University of Graz, 2012.

[4] D. Fuertinger, F. Kappel, S. Thijssen, N. Levin, and P. Kotanko. A model of erythropoiesis in adults with sufficient iron availability. *Journal of Mathematical Biology*, 66:1209–1240, 2013.

[5] L. Grüne and J. Pannek. *Nonlinear Model Predictive Control. Theory and Algorithms*. Springer, 2 edition, 2017.

[6] L. Lippmann. *Optimal administration strategies for EPO based on the model for erythropoiesis involving structured population equations with one structural*

variable. Diploma thesis, University of Konstanz, 2015. See http://nbn-resolving.de/urn:nbn:de:bsz:352-0-296098.

[7] K. Melcher. *Optimization with ODE Constraints for the Ultrafiltration Rate during Hemodialysis Treatment*. Master thesis, University of Konstanz, 2016.

[8] S. Rogg, D. Fuertinger, S. Volkwein, F. Kappel, and P. Kotanko. Optimal EPO dosing in hemodialysis patients using a nonlinear model predictive approach. *Journal of Mathematical Biology*, 79:2281–2313, 2019.

[9] J. Rohleff. *An Incremental Approach to Dynamic Mode Decomposition for Time-Varying Systems with Applications to a Model for Erythropoiesis*. Bachelor thesis, University of Konstanz, 2020. See http://nbn-resolving.de/urn:nbn:de:bsz:352-2-6fuu4fmo5tmo9.

Hardware tests for knee implants

WEAR TESTING OF KNEE IMPLANTS

Simulation and optimization

Hardware wear tests for knee implants, necessary for market authorization, are expensive and time-consuming. Faster than realtime wear simulations have the potential to speed up the design process for implants by providing frequent and early feedback. Computational support for virtual wear testing requires the detailed simulation of the mechanics of load cycles as well as the integration of resulting wear over many cycles. Moreover, sensitivity information helps to improve the implant's shape and to set up the hardware test correctly. Uncertainty quantification helps to judge reliability and variance of virtual and real test outcomes. The numerical simulation of implant wear, however, is computationally challenging and requires new algorithmic developments in order to be applicable in practice.

RALF KORNHUBER
Freie Universität Berlin

OLIVER SANDER
TU Dresden

ANTON SCHIELA
University of Bayreuth

MARTIN WEISER
Zuse Institute Berlin

PARTNERS

THOMAS BATSCH, aap Implantate AG, Berlin

CHRISTIAN ABICHT, Questmed GmbH, Kleinmachnow

Industrial challenge and motivation

The articular surfaces of total knee implants consist of a metallic femoral (upper) part and a tibial (lower) part usually made of high-density polyethylene. Due to daily motions, wear particles in particular from the tibial part accumulate in the joint and lead to inflammations, which is one of the limiting factors for the implant's life time. For marketing authorization, implant manufacturers have to demonstrate sufficiently low wear of their products by standardized wear tests.

Wear testing of total knee replacements is formalized by the ISO 14243 standard (Fig. 2). It requires to track the wear over a course of five million specified load cycles, and may require repeated manual intervention. A complete test takes about three months, at a cost of approximately 30 000 €. Both, time and cost, put a significant burden in particular on the design process. Faster than real time computational simulation of the wear rate can support the design process and reduce the time to market as well as the testing expenses.

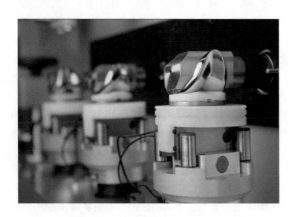

Figure 1: Hardware wear testing of knee implants. The femoral (upper) components undergo a prescribed motion for one to five million load cycles, after which the total wear is determined by the loss of mass in the tibial (lower) part.

The requirements formulated by the industry partners T. Batsch from aap Implantate AG, an implant manufacturer, and C. Abicht from Questmed GmbH, a testing laboratory, were, first, a quantitatively reliable and fast computation of total wear, and, second, the qualitatively correct reproduction of wear patterns on the articulating surfaces. A complete computational shape design was not required due a multitude of competing criteria in the design process. Instead, optimization of the freedom in the test setup left by the standard procedure as well as optimization-based hints for design improvements were desired.

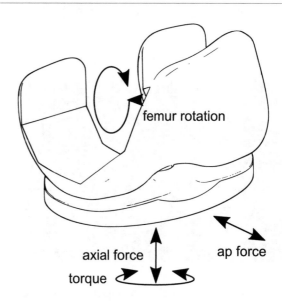

Figure 2: Load directions of the ISO 14243 standard.

Mathematical research

The computational support for wear testing has four aspects: (i) simulating the mechanics of single load cycles, (ii) computing the resulting wear over many cycles, (iii) determining the wear sensitivities with respect to the implant's shape and test setup, and (iv) the quantification of uncertainties. The problem size, in terms of required spatial resolution, test duration, and parametric variance, implies a huge computational effort that makes an industry-scale application of simulation challenging. New algorithmic developments led to viable simulation methods.

(i) Mechanical simulation. The simulation of a single load cycle requires the accurate solution of a sequence of mechanical contact problems. Since inertia is negligible compared to the applied forces, a quasi-static approximation is sufficient, and allows for the parallel computation of the displacement at all time steps. For such contact problems, robust and efficient non-smooth multigrid methods have been developed [2, 5]. From the resulting contact pressure and relative velocity of the sliding articular surfaces, the local wear rate can be computed by Archard's wear law [1]. Comparison with experimental data shows a very good agreement of the total wear rate, but some deviations in the wear pattern (Fig. 3).

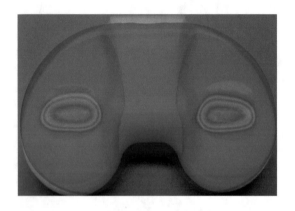

Figure 3: Overlay of simulated and experimental wear after 5 million gait cycles.

(ii) Long-time integration. The wear affects the implant's shape $\partial\Omega$ and thus the wear rate, but the high number of load cycles required by the ISO test prevents the simulation of every single cycle. Cycle jumping extrapolates the wear from a single cycle to a larger time interval, and permits to select a compromise between accuracy and simulation time (Fig. 4).

Essentially, it is equivalent to an explicit first order Euler time stepping for integration of the wear dynamics averaged over one load cycle of duration τ:

$$\dot{\overline{\partial\Omega}}(t) \approx \frac{1}{\tau} \int_{s=t}^{t+\tau} w(u(\overline{\partial\Omega}(t), s) \, ds$$

Higher-order time stepping schemes can, however, be much more efficient, in particular, if they exploit the accuracy-effort trade-off available in the mechanical simulation. Fast inexact spectral deferred correction (SDC) methods [4, 9] with

theoretically optimal selection of cycle simulation accuracy have been developed [10], and applied to implant wear, demonstrating a superior efficiency compared to simple cycle jumping, see Fig. 5. Both higher order and inexact evaluation of the right hand sides with adaptive accuracy increase the simulation efficiency significantly.

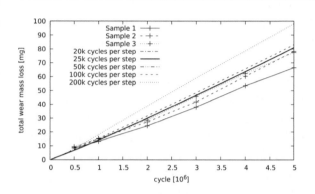

Figure 4: Total wear mass loss as a function of time for different cycle jumping step sizes.

(iii) Optimization of time-dependent contact problems. As a step towards the optimization of implant shapes, numerical methods for the optimal control of time-dependent contact problems were developed. The main difficulty is the non-smoothness of the problem, introduced by the contact constraint. This requires non-standard techniques, concerning analysis and optimization [3]. Time discretization techniques for dynamic contact problems were extended to the case of optimal control problems, optimality conditions and sensitivity results were derived, and numerical solution algorithms were implemented and tested [7, 8].

(iv) Uncertainty quantification. ISO wear tests are subject to subtle differences in environmental and initial conditions, such that their results have a certain distribution. Quantifying this uncertainty by sampling for mean increases the computational effort substantially. Therefore, adaptive hierarchical Monte Carlo methods have been developed [6], which reduce the sampling overhead significantly by exploiting again the accuracy-effort trade-off available in the mechanical simulation.

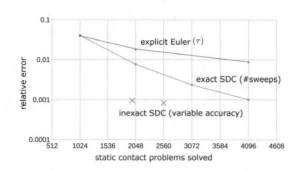

Figure 5: Efficiency of different time integrators for long-time wear simulation on 200 000 load cycles. The computational effort is measured by the number of static contact problems to be solved.

Implementation

In several mutual visits per year, both industry partners formulated their requirements and provided data in form of implant geometries and wear curves for validation as well as engineering experience and insight. The latter proved extremely helpful in formulating and addressing the mathematical problem in a practically relevant way. During the project, a prototype wear simulation software was implemented and validated in close collaboration with Questmed GmbH. Simulating wear patterns correctly turned out to be much more challenging than computing wear curves. This is mostly due to the particular stiffness structure of the wear dynamics, for which a preliminary treatment has been designed in the form of spatial wear rate smoothing approximating the action of an implicit integrator.

Industrial relevance and summary

The project consortium has successfully produced an efficient, massively parallel prototype code for the over-night simulation of wear during an ISO 14243 test. Simulated total mass loss and wear patterns were compared with experimental data provided by Questmed GmbH, with very good agreement. Moreover, aspects of shape and test optimization and treatment of uncertainties have been investigated as well. The approach is currently implemented by the

Algo4U Sagl, Lugano (https://algo4u.ch), for offering a commercial wear simulation service.

Acknowledgements

The authors acknowledge the support by the Federal Ministry of Education and Research (BMBF project no. 05M2013 (SOAK).

References

[1] J. Archard and W. Hirst. The wear of metals under unlubricated conditions. *Proc. Roy. Soc. London A*, 236(1206):397–410, 1956.

[2] A. Burchardt, C. Abicht, and O. Sander. Simulating wear on total knee replacements. Technical Report 1704.08307, arXiv, 2017.

[3] C. Christof and G. Müller. A note on the equivalence and the boundary behavior of a class of sobolev capacities. *GAMM-Mitt.*, 40(3):238–266, 2018.

[4] A. Dutt, L. Greengard, and V. Rokhlin. Spectral deferred correction methods for ordinary differential equations. *BIT Num. Math.*, 40(2):241–266, 2000.

[5] C. Gräser and O. Sander. Truncated nonsmooth newton multigrid methods for block separable minimization problems. *IMA J. Num. Anal.*, 39(1):454–481, 2018.

[6] R. Kornhuber and E. Youett. Adaptive multilevel monte carlo methods for stochastic variational inequalities. *SIAM J. Num. Anal.*, 56(4):1987–2007, 2018.

[7] G. Müller. *Optimal control of time-discretized contact problems*. PhD thesis, U Bayreuth, 2019.

[8] G. Müller and A. Schiela. On the control of time discretized dynamic contact problems. *Comp. Opt. Appl.*, 68(2):243–287, 2017.

[9] M. Weiser. Faster sdc convergence on non-equidistant grids by dirk sweeps. *BIT Numerical Analysis*, 55(4):1219–1241, 2015.

[10] M. Weiser and S. Ghosh. Theoretically optimal inexact spectral deferred correction methods. *Comm. Appl. Math. Comp. Sci.*, 13(1):53–86, 2018.

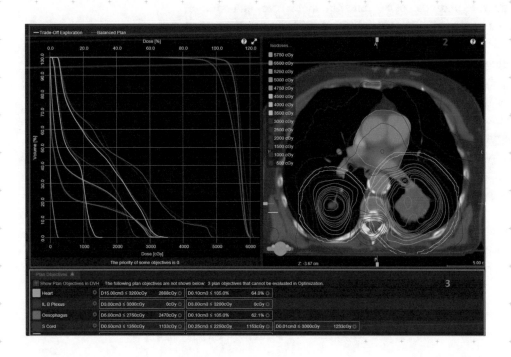

Interactive dose volume histogram and dose
distribution on patient anatomy

H. G. Bock et al. (eds.), *German Success Stories in Industrial Mathematics*,
Mathematics in Industry 35, https://doi.org/10.1007/978-3-030-81455-7_16

RADIOTHERAPY TREATMENT PLANNING

Interactive multi-criteria decision support

Radiotherapy treatment is one of the most important cancer therapies, as approximately half of all cancer patients are treated with radiation in some form during their treatment. Planning radiation treatment is challenging because the potential success of the the therapy has to be weighed against potential unwanted side effects, and treatment is highly individual: it is *a priori* unknown what the chances and limitations of the therapy are for a particular patient. There was a need for innovation in clinical treatment planning, as prior to our new planning technique, physicians would have to make a choice from multiple treatment plans produced by trial-and-error based on manually modifying priorities among the conflicting goals of the therapy.

It was by thorough mathematical analysis, development of new mathematics specific to the challenges in radiotherapy planning, and an innovative invention of a new graphical user interaction that Fraunhofer ITWM enabled its commercial partners a disruptive innovation in clinical radiotherapy treatment planning. Interactive multi-criteria treatment planning is the new gold standard, with potential to reduce planning time and, at the same time, improve plan quality by as much as 30% (amount of reduction of the dose in specific organs while retaining the curative potential of the plan).

KARL-HEINZ KÜFER
PHILIPP SÜSS
Fraunhofer Institute ITWM,
Kaiserslautern

PARTNERS

VARIAN MEDICAL SYSTEMS, Palo Alto, USA

Industrial challenge and motivation

With an estimated incidence number of 18.1 million world-wide in 2018 [5], cancer is one of the most severe health burdens on society today. Radiotherapy is one of the major therapy concepts for cancer: approximately half of all patients suffering from cancer are treated with radiotherapy at some point [2, 4]. State-of-the-art intensity modulated radiotherapy (IMRT) uses photon radiation from a linear accelerator and a collimator with movable leaves in order to deliver a dose that is precisely matched to the target. Other radiotherapy treatment techniques using photons and electrons exist, and some modern forms are based on protons and heavy ions.

Even with highly sophisticated technology, the challenge in radiotherapy is to realize a sufficiently high dose in the cancerous tissue while simultaneously sparing healthy tissue as much as possible. The first concern is to avoid side effects resulting from radiation damage to organs. The possibility of so-called "second cancer" caused by radiation treatment itself [1] has prompted physicians to pay even more attention than they had before to the dose received by the patient in the entire body.

Radiotherapy treatment planning is done using commercially available software - often included in the hardware that treatment sites purchased. Given the intricate balance of maximizing the curative (or palliative) effect on the one hand, and minimizing the risk of side effects on the other, treatment planning is a multi-criteria optimization problem. For the last two decades, researchers have worked to solve this problem both in theory and with regard to clinical practice.

Using software prior to first commercial implementations of a multi-criteria decision support tool, a planner would calculate many candidate treatment plans by manually altering priority weights for each clinical goal. This process would continue until the planner had gained a satisfactory treatment plan – a very subjective notion depending heavily on the planner's training as well as her intuition. The main problem with this approach is that the planner has to possess considerable expertise with the planning software to know just how the priorities of the goals would have to be manipulated to learn about the particular patient's chances and limitations for a therapy. In close collaboration with the German Cancer Research Center (DKFZ) at first, later with industrial partners and other scientists and hospitals, Fraunhofer ITWM set out to develop a system for treatment planning that would allow the planner to interact with the software based on her medical training instead of her software skills.

Mathematical research

The goal of multi-criteria radiotherapy planning is to create a plan that correspond best to prescribed clinical goals. To achieve this, a multi-criteria workflow begins with posing an optimization model. A planner specifies several objective functions and constraints for target structures and organs at risk that ideally reflect the clinical goals. All treatment planning systems implement fast dose calculations to map a given set of treatment plan parameters x, to a dose distribution $d(x)$, where x is typically a vector of several thousands entries, and $d(x)$ is a vector with around a million entries containing the radiation dose in the voxelized patient body. Typically, objective functions are based on dose statistics in regions of interest, e.g., dose-volume point (lower or upper) objectives defined on dose volume histograms (DVH) curves, generalized equivalent uniform dose (gEUD) [7] objectives, mean dose objectives, and DVH line objectives. Some objectives and constraints can address the machine parameters themselves; therefore, the objectives are a function of the parameters: $F : \mathbb{R}^n \mapsto \mathbb{R}^t$, that is, the t-dimensional vector of objectives is given by $F(x)$, and most of the objectives will contain the mapping of the dose as well: $F_i : \mathbb{R}^n \mapsto \mathbb{R}^m \mapsto \mathbb{R}$, where the inner mapping is the calculation of the dose $d(x)$. Sometimes, it is not possible to directly translate a clinical goal into a suitable objective function - in this case, a closely correlated substitute will be used. Also, not all aspects of the spatial dose distribution can be addressed at this stage. Local insufficiencies such as hot and cold spots are typically treated in an additional step at the end of the planning process.

The multi-objective optimization model is then given by

$$F(x) = (F_1(x), .., F_t(x)) \to \min$$
$$x \in X,$$

where X can include any number of restrictions on the machine parameters or the dose distribution and minimization is to be understood in the multi-criteria sense, which we motivate next. Once the optimization model is specified, a plan database is generated: a number of individual plans is calculated such the plans themselves and their interpolations approximate the range of feasible optimal compromises of the optimization model. This range is called the Pareto frontier $Par(F,X)$. A Pareto-optimal treatment plan has the property that none of the individual criteria can be improved while at least maintaining the others. Thus, lowering the dose in one organ of a Pareto-optimal plan can only be achieved if the dose in the tumor is also lowered, or at least one more organ receives a higher dose, for example. In the plan generation process, each plan is the result of a specific, iteratively determined scalarization of the multiple objectives. In case when X is a convex set, and all involved objectives F_i are convex functions in x, theses scalarizations are attained by weighting and summing all individual objectives. That is, numerically, the treatment planning system has to solve a series of problems of the form

$$\sum_{i=1}^{t} w_i F_i(x) \to \min$$
$$x \in X,$$

where the weights w_i are iteratively determined by an approximation scheme to approximate the Pareto frontier.

The approximation of the Pareto frontier is performed automatically without the decision-maker's involvement and can be carried out until a desired approximation quality is achieved (see [3, 8, 9] and references therein). In the methods developed by the ITWM, successive inner and outer approximations ($\mathcal{I}^j$ and $\mathcal{O}^j$, respectively) are constructed after the j-th scalarization problem has

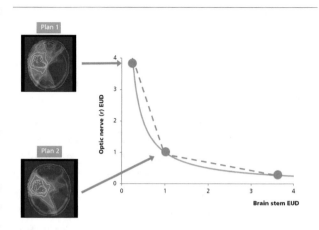

Figure 1: Illustrated navigation range for a hypothetical two-dimensional problem. The green dots are the three generated Pareto-optimal plans. They are on the Pareto frontier (orange line). The dotted line are the plans that are attainable by Pareto navigation.

been solved such that

$$\mathcal{I}^j \subseteq Par(F,X) \subseteq \mathcal{O}^j \quad \forall j$$

and the approximation quality is given by a conservative measure of how far the outer approximation differs from the inner approximation. The calculated plans are stored for navigation in the last stage, where they will be linearly interpolated (see figure 1). While the calculated database plans are Pareto-optimal, their interpolations are not; however, the distance to the nearest (hypothetical, not calculated) Pareto-optimal plan is limited by the approximation quality.

During the last stage of the workflow the decision-maker (DM) explores the generated inner approximation of the Pareto frontier. That is, the DM actually explores the plans of $\mathcal{I}^p$, after p plans have been calculated and the distance between $\mathcal{O}^p$ and $\mathcal{I}^p$ is sufficiently small. The DM specifies their preference by pulling a slider assigned to each of the objectives. The software calculates in real time optimal convex combinations of the previously calculated treatment plans and doses, which are displayed in real-time to the DM. By adjusting the objective combinations and receiving immediate graphical feedback, the DM can gradually steer towards the desired trade-offs [6].

Implementation

The first scientific collaboration between Fraunhofer ITWM and DKFZ began in 2000. Very early, ITWM identified the potential for the interactive Pareto navigation, culminating in the first patent [Setting control parameters, settings and technical parameters (Planning Guide). DE10318204B4] in 2001. Since then, the research funded by several public grants from the German Federal Ministry of Education and Research (BMBF) and the German "Krebshilfe" foundation, ITWM, DKFZ, and later the Massachussettes General Hospital (MGH), could demonstrate the practical and commercial potential of a multi-criteria decision-making framework. Several more patents (see list below) followed, strengthening ITWM's position to influence the commercial development of Pareto navigation in radiotherapy to this date. A project group at Fraunhofer ITWM had been working in close collaboration with R&D teams of Siemens Medical Solutions at first. However, Siemens decided to leave the market for radiotherapy treatment machines to focus on their medical imaging hardware in 2013. ITWM then granted RaySearch Laboratories, a software company for treatment planning the exclusive rights to the patents. They were the first to release a commercial implementation and continue to do so until today. In 2016, the patent rights were given to Varian Medical Systems, the world's largest supplier of radiotherapy treatment machines. Their treatment planning system Eclipse™ has the largest market share world-wide and features the multi-criteria decision-making tools described here. The collaboration with the commercial and medical partners to further innovation of interactive radiation treatment planning is ongoing.

Industrial relevance and summary

The introduction of interactive multi-criteria treatment planning was a game changer for the industry. It has to be noted, however, that its introduction by the way it was implemented in Varian's Eclipse™ in 2017 was not disruptive for the treatment planning process in clinics: care was taken to retain the user experience with the

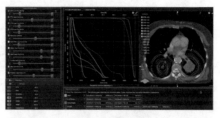

(a) Full screen

(b) Red box: Trade-Off exploration with slider bars for each selected objective.

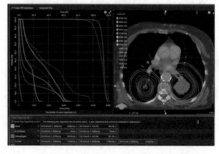

(c) Blue box: Graphical illustration of interactive dose volume histogram and dose distribution on patient anatomy. Green box: Plan objectives displaying the protocol constraints for dose volume monitoring.

Figure 2: Eclipse™ real time plan navigation screen view for a lung case.

"old" planning process and simply add the multi-criteria workflow on top, integrating it seamlessly with existing features of their software. As such, customer and user acceptation has been very high since the very first launch of what Varian calls the Multi-Criteria Optimization (MCO) tool. As Tianyou Xue, Ph.D., chief physicist at St. Luke's University Health Network in Pennsylvania notes: ".., the MCO tool is easy to use so we quickly learned how to balance target coverage with sparing healthy tissue, to create excellent plans for our patients."[10] Clinicians at the Beatson West of Scotland Cancer Center in Glasgow, U.K., also note how MCO has improved the quality of their planning across a wide range of treatment sites. They report "consistently seeing dramatic reductions in the dose to OARs—sometimes as much as 30% – depending on the plan" [10].

ITWM's strategy to supply the industry with innovative technology that users can adopt easily and with great benefit is a winning one, as these testimonies show. Starting by analyzing the mathematical problems of radiation treatment planning, and inventing new mathematical methods targeted to a high usability for a human decision-maker, it was possible to successfully introduce a disruptive innovation into the market.

The authors acknowledge the support by the Federal Ministry of Education and Research (BMBF project no. 01AK941, RADIOPLAN), (BMBF project no. 01B08002D, DOT-MOBI), (BMBF project no. 01B13001C, SPARTA)

Patents

» Küfer, Trinkaus, DE10318204B4

» Küfer et al., DE102010062079B4

» Küfer et al., US9824187B2

» Teichert, US20170239492A1

References

[1] American Cancer Society. How does radiation therapy affect the risk of second cancers? https://www.cancer.org/treatment/treatments-and-side-effects/physical-side-effects/second-cancers-in-adults/radiation-therapy.html, 2014. Accessed: 2019-03-20.

[2] R. Atun, D. A. Jaffray, M. B. Barton, F. Bray, M. Baumann, B. Vikram, T. P. Hanna, F. M. Knaul, Y. Lievens, T. Y. M. Lui, M. Milosevic, B. O'Sullivan, D. L. Rodin, E. Rosenblatt, J. V. Dyk, M. L. Yap, E. Zubizarreta, and M. Gospodarowicz. Expanding global access to radiotherapy. The Lancet Oncology, 10(16):1153–1186, 2015.

[3] D. Craft. Multi-criteria optimization methods in radiation therapy planning: a review of technologies and directions. arXiv:1305.1546v1, 2013.

[4] G. Delaney and M. Barton. Evidence-based estimates of the demand for radiotherapy. Clinical Oncology, 2(27):70–76, 2015.

[5] International Agency for Research on Cancer – WHO. All cancers fact sheet. http://gco.iarc.fr/today/data/factsheets/cancers/39-All-cancers-fact-sheet.pdf, 2018. Accessed: 2019-03-20.

[6] M. Monz. Pareto Navigation - Interactive multiobjective optimisation and its application in radiotherapy planning. PhD thesis, Technische Universität Kaiserslautern, 2006.

[7] A. Niemierko. A generalized concept of equivalent uniform dose. Medical Physics, (26):1100, 1999.

[8] J. Serna. Approximating the Nondominated Set of R+ convex Bodies. Master's thesis, Technische Universität Kaiserslautern, 2008.

[9] J. Serna, M. Monz, K. Küfer, and C. Thieke. Trade-off bounds for the Pareto surface approximation in multi-criteria IMRT planning. Physics in Medicine and Biology, 54(20):6299–6311, 2009.

[10] Varian Medical Systems. Multi-criteria optimization: Creating high-quality treatment plans in a fraction of the time. https://www.varian.com/oncology/events-resources/centerline/multi-criteria-optimization-mco-creating-high-quality, 2019. Accessed: 2019-03-20.

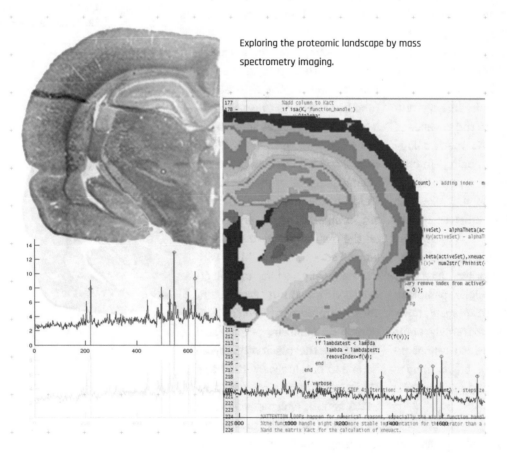

Exploring the proteomic landscape by mass spectrometry imaging.

MALDI IMAGING: EXPLORING THE MOLECULAR LANDSCAPE

Machine learning approaches for blind source separation of high dimension mass spectrometry data

The driving force behind recent developments in Life Sciences such as drug discovery, individual therapy planning or pathway detection in systems biology are the Omics-technologies (Proteomics, Lipidomics, Metabolomics, etc.). Over the last 15 years these technologies have been partially revolutionized due to the advance of a new bioanalytic methodology called MALDI imaging (matrix assisted laser desorption ionization). MALDI imaging allows to determine the complete molecular landscape of tissue sections and in combination provides the basis for biomarker detection, tumor typing or digital pathology in general. MALDI imaging, however, produces extremely complex data sets, which cannot be evaluated without bioinformatics tools for extracting task specific information.

These developments would not have been possible without the advance of novel mathematical theories for determining characteristic spectral patterns, which are the building blocks for classification schemes (tumor typing) or proteomic identification (biomarker detection, pathways).

The resulting mathematical algorithms are based on theoretical research at the interface of functional analysis (inverse problems) and numerical linear algebra (matrix factorization). These challenges are the core of the research activities of the Center for Industrial Mathematics, University Bremen. They have provided the mathematical foundations as well as – in collaboration with the nationally leading pathological service institute, Proteopath, Trier – prototypical implementations for supporting pathological diagnosis. The resulting publications and patents were the basis for founding a spin-off company, SCiLS GmbH, which developed a commercial software toolbox for analyzing MALDI imaging data. This company developed into the worldwide leader in this fields with presently more then 80 % marketshare worldwide.

PETER MAASS
LENA HAUBERG-LOTTE
TOBIAS BOSKAMP
Center for Industrial Mathematics, University of Bremen

PARTNERS

PROF. DR. DR. JÖRG KRIEGSMANN, **Proteopath GmbH, Trier** | DR. DENNIS TREDE, **Bruker Daltonik GmbH, Bremen**

Industrial challenge and motivation

Mass spectrometry imaging (MSI), in particular matrix-assisted laser desorption/ ionization (MALDI) MSI, is a label-free technique for spatially resolved molecular analysis of biological tissue samples (molecular landscape), with a broad range of applications in pharmaceutical and biomedical research [12].

This is a most remarkable development for a comparatively novel technology, which started in 2005 with the first commercially available MALDI imaging devices. Since then, it has become a routine technique for all kinds of applications in proteomics, lipidomics or metabolomics as a basis for biomarker detection, drug discovery or pathway analysis in systems biology [5, 11, 10].

With recent technological advances in acquisition speed and robustness, applications of MALDI MSI in pathological diagnostics became feasible, where this method can help to, for example, characterize tumor tissue, delineate tumor regions or identify tumor subtypes [6, 8]. Future developments aim at establishing such tools for digital pathology in clinical routine applications for supporting tumor diagnosis and individual therapy planning.

This success is based only partially on a combination of technological developments for building the mass-spectrometric hardware and recent bio-analytic developments in proteomics. In addition, analyzing the huge data set produced by MALDI imaging experiments and the task to extract diagnostically meaningful information, poses an enormous challenge in mathematics and bioinformatics.

In MALDI MSI, a pulsed laser beam is focused at a series of spots (pixels) covering a tissue sample with a spatial resolution ranging from 5 to 250 micrometer. Biomolecules over a wide mass range are desorbed from the tissue surface, ionized and fed into a mass spectrometer. The resulting dataset consists of a complete mass spectrum with up to 1 million spectral intensity values for each tissue spot, yielding extremely high-dimensional multispectral images for arbitrary ion masses.

The overwhelming richness of this data set requires new bioinformatics tools for extracting relevant information, which is stable with respect to the

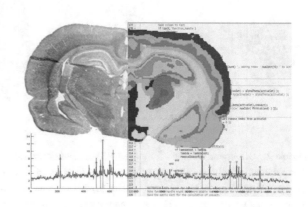

Figure 1: Anatomical section of a rat brain and segmentation based on MALDI spectral fingerprints

individual biological diversity of human metabolisms but also with respect to the sensitive sample preparation and measurement procedures [9], see Fig. 2. The core problem asks to segment spatial tissue regions with different biological characteristics, i.e. the identification of spectral fingerprints associated with different tissue phenotypes.

Mathematical research

A MALDI imaging data set consists of mass spectra, which are measured at different locations (pixels) of a tissue section. This yields a hyperspectral data set consisting of several thousands of channels (m/z mass spec values) at every pixel. For training purposes we assume, that MALDI data from several tissue sections are available together with an expert annotation indicating regions with different metabolic structures, e.g. discriminating between tumorous and non-tumorous regions. The task is to determine **spectral fingerprints** 1 such that correlating a mass spectrum at a certain position with these fingerprints yields a feature vector, which allows a straight forward classification or tumor typing.

Spectral fingerprints of such protein biomarkers typically consist of a characteristic pattern of spectral peaks corresponding to different peptides, as well

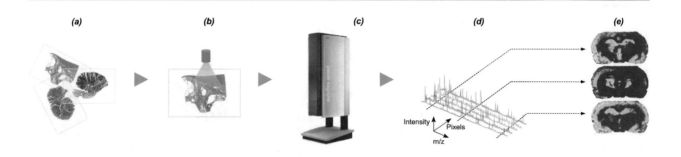

Figure 2: Schematic workflow for MALDI MSI of FFPE tissue samples. Tissue sections (a) are subjected to sample preparation including deparaffination, antigen retrieval, on-tissue tryptic digestion, and matrix application (b). Prepared tissue sections are inserted into the MALDI MSI instrument (c) and mass spectra (d) are acquired. When fixing single m/z-values the intensities in the measurement area can be visualized as m/z-images (e), reflecting the molecular distribution of peptides with corresponding masses.

as molecular modifications and isotopes. This observation, together with the fact that spectral intensities are strictly non-negative, motivates the use of non-negative matrix factorization (NMF) algorithms for analyzing MALDI MSI datasets.

This approach is based on the underlying assumption, that different sections of tissue sections can be characterised by their different metabolic processes and that these different metabolic processes lead to distinctly different molecular structures.

In a mathematical formulation, MALDI imaging provides a data matrix which consists of data vectors $Y_{i,\cdot} \in \mathbb{R}_{\geq 0}^m, i = 1, \ldots n$, where the index i refers to a spatial position and $Y_{i,\cdot}$ represents the spectrum measured at this position. The assumption that only a small number p of metabolic processes or molecular structures are represented in the data set, motivates that p spectral patterns $X_{k,\cdot} \in \mathbb{R}_{\geq 0}^m, k = 1, \ldots, p$ with $p \ll \min(n, m)$ are sufficient for approximating the full data set, and that there exist coefficients $K_{ik} \in \mathbb{R}_{\geq 0}$ such that $Y_{i,\cdot} \approx \sum_{k=1}^p K_{ik} X_{k,\cdot}$. This results in a low-rank approximation of the data matrix Y by non-negative pseudo spectra $X_{k,\cdot}$, the rows of X, and non-negative pseudo channels $K_{\cdot,k}$. Hence, the mathematical task is to determine a non-negative matrix factorization (NMF) of Y by

$$\min_{X,K} \frac{1}{2} \|Y - XK\|^2 .$$

This seemingly simple problem in numerical linear algebra is severly ill-posed and requires specific stabilization techniques. Moreover, for large data sets and for incorporating the spatial coherence with respect to the positions, where the spectra are measured, one should consider the continuous limit. This opens a connection to functional analytic regularization theory, which aims at analyzing

$$\min_{X,K} \frac{1}{2} \|Y - XK\|^2 + \alpha \Phi(X) + \beta \Psi(K)$$

in a function space setting. We have analyzed such regularization scheme analytically, which leads to novel iterative algorithms for determining NMF-decompositions for large data sets, [7, 4].

Based on these spectral pattern one can then attack the construction of classification schemes for tumor typing. Classical approaches, based on this two step procdure of first determining spectral patterns and then designing classification schemes, have been widely investigated, see [4]. These methods typically utilize statistical tests for detecting discriminative spectral features, such as principal component analysis (PCA), probabilistic latent semantic analysis (PLSA) or non-negative matrix factorization (NMF) in the first step. These spectral features then form the basis for constructing a subsequent classification scheme (LDA, logistic regression, etc.), [2, 3].

This process is greatly enhanced by incorporating the classification model directly into the matrix factorization scheme. Let us assume, that - for

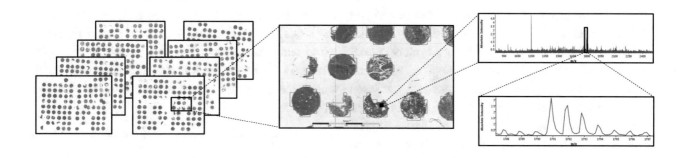

Figure 3: Schematic data structure. Optical images of 8 hematoxylin and eosin (HE) stained TMAs (left), each containing multiple core biopsies of lung cancer patients (middle). For each core several MALDI TOF spectra are collected (upper right). The close-up shows an isotopic pattern for a peptide at $m/z 1790.9$ (lower right).

training of the classification scheme - a pathological expert has marked different tissue regions by 0 (healthy) or 1 (tumor). One standard procedure is to approximate these given binary class labels $u \in \{0,1\}^n$ by a linear combination of the above described correlations $Y X_{k,\cdot}{}^t$. Thus, coefficients $\beta \in \mathbb{R}^p$ have to be determined, such that $u \approx \sum_{k=1}^p \beta_k Y X_{k,\cdot}{}^t = Y X^t \beta$. The estimation of β is typically done with a least squares model and leads to the minimization problem

$$\min_\beta \|u - Y X^t \beta\|_F^2 \ . \tag{1}$$

The minimization using the available annotated training data yields a suitable parameter set $\hat\beta$ and a corresponding characteristic vector $\hat{x} = X^t \hat\beta$. Better results are obtained by logistic regression models and specially adapted minimization schemes.

On the applied side we have analyzed the convergence properties of different minimization schemes and we have applied these methods to clinical data obtained for several hundred patients [1], see Fig. 4 as well as the list of patents below.

Implementation

The industrial side of our MALDI research started with a patent [Verfahren zum rechnergestützten Verarbeiten von räumlich aufgelösten Hyperspektraldaten, insbesondere von Massenspektrometriedaten. DE102013207402A1], which was sold to Bruker Daltonik, the world leading manufacturer of MALDI imaging hardware. As scientists we were eager to see our results being implemented in the 'real' world. However, two years later, nothing had happened, there was no visible sign, that Bruker would implement the patent and use it as part of their software. We asked Bruker, why they didn't use our results and got the answer, that is was interesting but too complicated, they would not use it unless we would provide it in directly usable software.

This was the initial spark for founding a spin-off, SCiLS (scientific computing in life sciences). The first two years were supported by the local business development fund of the City of Bremen (WFB project no. FUE485B, MALDI-Imaging Basisexperiment), which allowed us not only to implement the patent but also to re-implemented and to improve considerably the existing software for analyzing MALDI imaging data. It definitely helped, that this was and still is a rather young technology, the first commercial MALDI imaging hardware was build in 2005 and, hence, the software part was still very much in its initial state. Also, the Bremen side of Bruker Daltonik is still a company very much focussed on hardware and biochemistry, with limited resources for software development.

This led to an intense collaboration between pathological experts from Proteopath GmbH, Trier, (Prof. Dr. Joerg Kriegsmann), who in collaboration with the Institute of Pathology, University Hospital Heidelberg, provided, MALDI data as well as expert annotation for several hundreds of patients, ZeTeM, University of Bremen, which developed the bioinformatics tools for data processing and SCiLS GmbH, which developed the software toolbox SCiLS Lab specialized on MALDI imaging. Research on the challenging theoretical questions as well as prototypical algorithmic developments were supported by two projects (MaDiPath, Digital Staining). This resulted in a software, which was finally licensed by Bruker Daltonik. SCiLS quickly developed to become the leading supplier for mass spectrometric imaging software and our small company received an Enterprise Award as the best start-up in Bremen in 2015. SCiLS was acquired by Bruker in 2017, with our former PostDoc as CEO and about five of our former students working in the company.

Industrial relevance and Summary

MALDI Imaging has become a standard bioanalytic procedure in pharmaceutical and medical research. Present developments aims at implementing MALDI imaging based procedures in clinical routine applications in particular for supporting tumor diagnostics. The mathematical research described in this section is relevant for the short as well as the long- term development of digital pathology.

Clinical routine applications for such automated diagnostic tools will still require several years before becoming reality. This most likely will be based on a mix of different bioanalytic techniques for proteomic and metabolomic analysis. Presently, MALDI imaging is a most promising candidate to do its potential for obtaining complete and label free information of the molecular landscape of tissue sections. The sheer amount of information contained in a single data set requires computer-aided techniques for any type of information extraction. This is exactly were the presented mathematical tools for extracting diagnostically relevant spectral patterns kicks in 4.

On the short term, the work on the theoretical side of the project, which was supported by the projects 'MaDiPath' and 'Digital Staining' has stimulated subsequent experimental projects, which lead to further publications and patents, which have been included in the SCiLS Lab software of our co-

operation partners. This software, which started to be developed by a start-up company of the center for industrial mathematics in Bremen, is now the world wide standard in the field of MALDI imaging. The software has a market share of more than 80 % worldwide with more than a thousand installations worldwide.

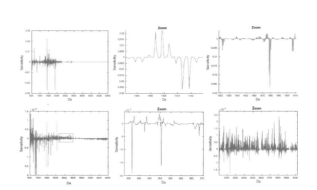

Figure 4: Sensitivity of IsotopeNet on task ADSQ, where the sensitivity was computed for the predicted probability of class AD (upper left). Zoom in with high peaks at 1406.6 and 1410.7 Da upper middle). Zoom in with right peaks at 1821.8, 1877.8 and 1905.9 Da (upper right). Sensitivity of IsotopeNet task LP, where the sensitivity was computed for the predicted probability of class Lung (lower left). Zoom in with high peaks at 836.5, 852.4 and 868.5 Da (lower middle). Zoom in showing in high oscillations in the range of 2100 − 2900 Da (lower right).

Acknowledments

The authors acknowledge the support by the Federal Ministry of Education and Research (BMBF KMU-innovativ project no. 13GW0081C, MaDiPath), the Federal Ministry for Economic Affairs and Energy (BMWI ZIM project no. KF2443904SB4, Digital Staining) and the Bremeninvest (WFB project no. FUE0485B, MALDI-Imaging Basisexeperiment).

Patents

» T. Boskamp, D. Lachmund, DE 10 2017 008 885.3, 22.09.2017. US Patent and Trademark Office 16/135,496, 19.9.2018.

» T. Boskamp, DE 10 2016 012 302 A1 and PCT Application PCT/EP 2017/001131, 2017

» P. Maass, et al.   DE102014224916A1 US 20160163523 2015

» P. Maass, et al.   DE102013207402A1, 24.04.2013, issued 30.10.2014.

» D. Trede, P. Maass, F. Alexandrov: DE10 2011 003 242.8, 2012

» D. Trede, P. Maass, H. Preckel, Europäisches Patentamt   EP2128815,   07.05.2008, US2011/0098198 A1 2011.

References

[1] J. Behrmann, C. Etmann, T. Boskamp, R. Casadonte, J. Kriegsmann, and P. Maass. Deep learning for tumor classification in imaging mass spectrometry. *Bioinformatics*, 34(7):1215–1223, 2017.

[2] T. Boskamp, D. Lachmund, J. Oetjen, Y. C. Hernandez, D. Trede, P. Maass, R. Casadonte, J. Kriegsmann, A. Warth, H. Dienemann, et al. A new classification method for maldi imaging mass spectrometry data acquired on formalin-fixed paraffin-embedded tissue samples.   *Biochimica et Biophysica Acta (BBA)-Proteins and Proteomics*, 1865(7):916–926, 2017.

[3] Y. Cordero Hernandez, T. Boskamp, R. Casadonte, L. Hauberg-Lotte, J. Oetjen, D. Lachmund, A. Peter, D. Trede, K. Kriegsmann, M. Kriegsmann, et al. Targeted feature extraction in maldi mass spectrometry imaging to discriminate proteomic profiles of breast and ovarian cancer. *PROTEOMICS–Clinical Applications*, page 1700168, 2018.

[4] P. Fernsel and P. Maass.   A survey on surrogate approaches to non-negative matrix factorization. *Vietnam Journal of Mathematics*, 46(4):987–1021, 2018.

[5] M. R. Groseclose, P. P. Massion, P. Chaurand, and R. M. Caprioli. High-throughput proteomic analysis of formalin-fixed paraffin-embedded tissue microarrays using maldi imaging mass spectrometry. *Proteomics*, 8(18):3715–3724, 2008.

[6] J. Kriegsmann, M. Kriegsmann, and R. Casadonte. Maldi tof imaging mass spectrometry in clinical pathology: a valuable tool for cancer diagnostics. *International journal of oncology*, 46(3):893–906, 2015.

[7] J. Leuschner, M. Schmidt, P. Fernsel, D. Lachmund, T. Boskamp, and P. Maass. Supervised non-negative matrix factorization methods for maldi imaging applications. *Bioinformatics*, 2018.

[8] R. Longuespée, R. Casadonte, M. Kriegsmann, C. Pottier, G. Picard de Muller, P. Delvenne, J. Kriegsmann, and E. De Pauw.   Maldi mass spectrometry imaging: A cutting-edge tool for fundamental and clinical histopathology. *PROTEOMICS–Clinical Applications*, 10(7):701–719, 2016.

[9] J. Oetjen, D. Lachmund, A. Palmer, T. Alexandrov, M. Becker, T. Boskamp, and P. Maass. An approach to optimize sample preparation for maldi imaging ms of ffpe sections using fractional factorial design of experiments. *Analytical and bioanalytical chemistry*, 408(24):6729–6740, 2016.

[10] J. Oetjen, K. Veselkov, J. Watrous, J. McKenzie, M. Becker, L. Hauberg-Lotte, J. Kobarg, N. Strittmatter, A. Mróz, F. Hoffmann, et al. Benchmark datasets for 3d maldi-and desi-imaging mass spectrometry. gigascience 4: 20, 2015.

[11] J. Quanico, L. Hauberg-Lotte, S. Devaux, Z. Laouby, C. Meriaux, A. Raffo-Romero, M. Rose, L. Westerheide, J. Vehmeyer, F. Rodet, et al. 3d maldi mass spectrometry imaging reveals specific localization of long-chain acylcarnitines within a 10-day time window of spinal cord injury. *Scientific reports*, 8(1):16083, 2018.

[12] E. H. Seeley and R. M. Caprioli. Maldi imaging mass spectrometry of human tissue: method challenges and clinical perspectives. *Trends in biotechnology*, 29(3):136–143, 2011.

An underground station in Nürnberg

ENERGY-EFFICIENT TIMETABLING IN A GERMAN UNDERGROUND SYSTEM

Discrete optimization at the interface of logistics and energy management

Timetabling of railway traffic and other modes of transport is among the most prominent applications of discrete optimization in practice. However, it has only been recently that the connection between timetabling and energy consumption has been studied more extensively. In our joint project with VAG Verkehrs-Aktiengesellschaft, the transit authority and operator of underground transport in the German city of Nürnberg, we develop algorithms for optimal timetabling to minimize the energy consumption of the trains. This effect shall be achieved via more energy-efficient driving as well as increasing the usability of recuperated energy from braking. Together with VAG, we have worked extensively to establish a broad basis of operational data, for example characteristic power consumption profiles as well as travel time and dwell time distributions for the trains running in the network, to serve as input to our optimization methods. On the collected data sets, our approach has already shown significant potential to reduce energy consumption and, as a consequence, electricity costs and environmental impact. Furthermore, mathematical analysis of the polyhedral and graph structures involved in the optimization approach have enabled us to compute high-quality solutions within short time. This positive outlook motivated VAG to extend this project to include further operational constraints in the model and to adopt the resulting software planning tool in practice afterwards. It will assist timetable planners at VAG in using the available degrees of freedom in their timetable drafts to optimize the energy-efficiency of the underground system.

ANDREAS BÄRMANN
Friedrich-Alexander University
Erlangen-Nürnberg

PATRICK GEMANDER
Fraunhofer Institute for Integrated
Circuits IIS

ALEXANDER MARTIN
Friedrich-Alexander University
Erlangen-Nürnberg,
Fraunhofer Institute for Integrated
Curcuits IIS

MAXIMILIAN MERKERT
Otto von Guericke University Magdeburg

PARTNERS

FREDERIK NÖTH, **VAG Verkehrs-Aktiengesellschaft**

Industrial challenge and motivation

Traction energy consumption is among the most important cost factors in the electricity bill of a railway undertaking. It is greatly influenced by the manner in which the trains are driven. Thus, a significant reduction in energy consumption can be achieved by choosing energy-efficient velocity profiles. This includes making use of pure rolling phases, the so-called *coasting*, as far as possible, as a train consumes no traction energy at all in this phase and, due to the low rolling friction, only slowly looses speed.

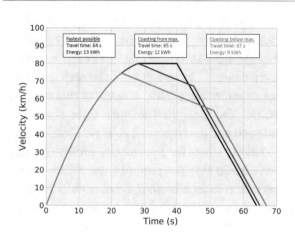

Figure 2: Schematic representation of the effect of choosing between one of the three driving modes "always driving as fast as possible" (*black*), "accelerating to maximum velocity followed by coasting" (*blue*) and "accelerating to below maximum velocity followed by extended coasting" (*red*) on a sample underground leg

Figure 1: An underground station in Nürnberg, ©VAG – Claus Felix

In Figure 2, the effects of choosing between different driving modes of a train are shown schematically for a train in an underground network. These data show that by slightly slowing down the fastest possible speed profile on a given track, the train may consume up to 1/3 less in energy. In this respect, it is especially beneficial to extend the coasting phases of the train as much as possible. Altogether, choosing the optimal velocity profile for each train on each leg (= timetabled run between two stations) with respect to given total line travel times entails a huge leverage for bringing down the consumption of the overall underground system. This finding motivated our joint research project with *VAG Verkehrs-Aktiengesellschaft*, the local operator of public transport in the German city of Nürnberg. Its idea was to take a given timetable draft towards

the end of the timetable planning phase and to use the remaining degrees of freedom to slightly shift train departures within fixed windows around their currently planned departure times. The aim is to create the necessary flexibility to enable choosing the best possible velocity profile on each leg. At the same time, these shifts in the departures times allow for a better synchronization of departure and arrival events. This is important as a braking train is able to feed back recuperated energy to the grid. However, this energy can only be used if there is another train in the network which is accelerating at the same time, otherwise it is lost. Overall, there is a considerable potential for cost saving, as we have demonstrated in our collaboration. In the following, we will elaborate on the mathematical approach and present our case study for the underground system of Nürnberg.

Mathematical research

Based on a timetable draft created by expert planners, the studied task is to determine slight modifications in the train departure times as well as choosing velocity profiles for all trains in an energy-optimal way. However, these modifications shall retain the timetable structure established in the draft according to stated criteria, e.g. *dwell times* (= passenger interchange times) in the stations, *minimum headway times* (= safety distances) between trains and desired connections between trains. In order to construct a mixed-integer programming (MIP) model for this task, we discretized the time horizon into time steps of 1 second each and determined a suitable discrete set of (e.g. 3) alternatives for the velocity profiles for each train on a given leg. The profiles were initially chosen as heuristic solutions to an optimal control problem; later we changed them against measured profiles of actual train runs in the network. Furthermore, we allowed departure time shifts up to a given amount, e.g. 15 seconds around the draft departure time for each leg – a change that is hardly noticeable by the passengers but that can still allow for significant energy savings as we were able to show. With allowed shifts of ± 15 seconds in increments of 5 seconds and 3 profiles to choose from, there are already $3 \cdot 7 = 21$ possible choices for the combination of departure time and velocity profile for each leg. Given that there are $24,000$ legs to be served in the Nürnberg underground each day, this means there are $21^{24,000} \approx 10^{31,733}$ possible timetable adjustments to choose from. No company planner could hope to evaluate all of them manually in order to determine the most energy-efficient one. Via the techniques of discrete optimization we have developed over the course of this project, however, we are able to produce near-optimal timetables within one hour or less.

To this end, we came up with a model formulation for the set of the feasible timetable adjustments as a special case of the clique problem with multiple-choice constraints on an undirected graph G (see [1, 2]). Its nodes represent possible combinations of departure time and velocity profile for the legs to be scheduled, while the edges model compatibilities between the departure configurations for different legs. Whenever the departure configurations for two specific legs do not violate any requirements for a feasible timetable, such as the above-mentioned ones, the corresponding nodes are connected by an edge. This results in an optimization model of the type

$$\min \quad \sum_{t \in T} \max(P(x,t), 0)$$
$$\text{s.t.} \quad x \in X,$$

where $P(x,t)$ represents the total energy consumption at time step t, summed over all running trains, while X is the set of feasible timetable adjustments. Taking the maximum of $P(x,t)$ and 0 reflects that energy from a braking train can only be recuperated if it is used by other trains in the same time step. After linearizing the objective function with the help of additional auxiliary variables, the above model can be written as an MIP. We point out that all relevant types of timetabling constraints can indeed be expressed as pairwise node conflicts, which constitutes a very special structure. There are several ways to translate them into linear constraints. However, modelling the feasible region X in the most efficient way is very important as standard MIP solvers cannot solve the problem efficiently for real-world networks if a naive model formulation is used.

Our search for an adequate model formulation was inspired by [5]. Its authors were among the first to study the combined optimization of railway (or more precisely underground) timetables and energy consumption, giving a heuristic for reducing instantaneous power peaks. In [3], we took up their basic idea and studied the effects of optimal timetabling for small subnetworks of German railway traffic under different objective functions relating to power consumption patterns. During this work, we realized that the problem contains an interesting structure to be exploited in order to reduce solution times. The nodes of the compatibility graph can be partitioned by the legs they belong to, and within each partition V_l they can be sorted by departure time. For the special (but still NP-hard) case of a single energy profile available for each leg, the compatibility structure then allows for a totally unimodular description of the timetabling polytope. It could be improved to an even more efficient dual-flow formulation by using the canonical ordering of the departure times for each leg. This special structure also comprises problems in other application contexts, such as the piecewise linearization of path flows – e.g. of natural gas in a pipeline (see [4]). We generalized the core properties of the compatibility structure to the abstract notion of *staircase compatibility* in [1]. There, the resulting model formulations were successfully employed on much larger subnetworks of *Deutsche Bahn AG (DB)*, up to the Germany-wide network, for minimizing peak power consumption. When using our improved formulations, we observed significant savings in solution times (over a factor of 100 in several cases), which allowed us to solve the problem for the Germany-wide network within a couple of minutes, but took hours to solve beforehand.

For multiple energy profiles per leg to choose from, the structure of the feasible set still tends to favour similar reformulations but cannot be perfectly described by staircase compatibility. We continued our polyhedral studies by considering a special case with respect to the *dependency graph* of the subsets in the node partition according to legs. This is the graph that encodes which pairs of subsets of nodes directly impose *any* restrictions on each other. We showed that the feasible set in this case can be completely described by stable-set inequalities if the dependency graph is cycle-free. This leads to the following overall formulation:

$$\min \quad \sum_{t \in T} \max(P(x,t),0)$$

$$\text{s.t.} \quad \sum_{v \in V_l} x_v \;=\; 1 \qquad \forall \text{ subsets } V_l$$

$$\sum_{v \in S} x_v \;\leq\; 1 \qquad \forall \text{ stable sets } S \text{ in } G$$

$$x \;\in\; \{0,1\}$$

where $G = (V,E)$ is the compatibility graph and the subsets V_l for each leg l form a partition of V. Altogether, we want to choose exactly one departure configuration for each leg, as modelled by the variables x_v for each node $v \in V$. The stable sets in G represent exactly the subsets of nodes which are in pairwise conflict with each other. Note that the total number of stable sets is potentially large and difficult to generate in general. However, only stable sets involving nodes from just two subsets are needed, which significantly reduces the enumeration effort and the size of the formulation – especially if the number of departure configuration choices for each leg is small in comparison to the number of legs in the timetable.

We used this improved formulation to greatly reduce solution time for optimizing the timetable in the *Nürnberg* underground network, see [2], also for more details on the aforementioned polyhedral results. In this preliminary computational study, an optimized shifted schedule of the longest line *U1* in the system reduced the overall energy consumption by about 18% during the morning rush hour interval between 5 a.m. and 9 a.m. when compared to the actual 2018 schedule. From there on, we have undertaken great efforts to broaden the available data base and to extend the results to all three Nürnberg underground lines over the whole day in order to see how much of these 18% in savings can be expected to be obtained in practice. These efforts and the findings we had will be described in the next two sections.

Implementation

The initial spark for this project was a cooperation with DB in project *E-Motion* (2013–2016), funded by the German Ministry of Education and Research (BMBF). Its aim was to come up with optimization algorithms to compute slight adjustments in the departure times of the trains to reduce peak power consumption for railway transport. The successful completion of this project led us to approach VAG in order to see if the same technique could be used to reduce peak consumption in the Nürnberg underground system. We soon learned that an even higher potential lies in reducing the overall power consumption by choosing energy-efficient driving patterns. So whereas in the project with DB we only adjusted departure times, the new task was to choose optimal travel times for each leg. After extending our timetabling model accordingly, we iteratively increased its performance. Firstly, the new degree of freedom, choosing travel times, added complexity to the mathematical model. An extensive study of its structure as described in [2] enabled us to give a more compact problem formulation that was much easier to solve and still respected all necessary constraints. Secondly, we incorporated

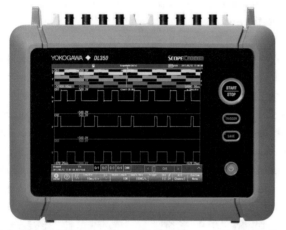

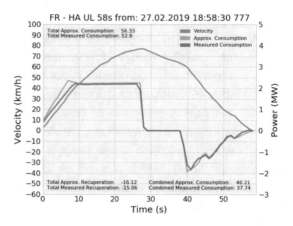

(a) The DL350 device for measuring the power consumption of a train, i.e. the traction current used by the power trains (©Yokogawa Deutschland GmbH)

(b) A sample velocity profile (*blue*), approximated (*red*) and measured (*green*) power consumption profile on a given leg in the Nürnberg underground network

Figure 3: The company VAG has ordered and installed a dedicated device that measures the power consuption of a train running in the Nürnberg underground at a resolution of 200 ms. This allows for the derivation of precise velocity and power profiles as input to our timetable optimization model. As it turned out, the approximations of power consumption based on velocity and acceleration we used before were quite accurate in most cases.

additional timetabling constraints to make sure that our solutions are not just energy efficient, but also real-world applicable. Note that these additions seamlessly fit into our new mathematical framework and therefore had no negative impact on solution times. Finally, we replaced the simulated velocity and power profiles, which were based on theoretical knowledge of train characteristics, and which we used in the beginning, with profiles that were based on actual measurements for train runs in the system. For some time, we used profiles approximated from velocity and acceleration measurements as recorded by the tachographs in the wagons together with characteristic power consumption curves of the power trains. After confirming the potential of our approach on these approximate power profiles, VAG purchased and installed a dedicated device for measuring the traction current used by the powertrains in a wagon. Figure 3 shows this DL350 device as well as a sample profile recorded by it.

These more precise recordings improved the accuracy of the model output further, and, as our database of sample power profiles grew, we could cooperate with the timetable experts

at VAG on a related topic as well. Namely, the collected data allowed us to perform broad statistical evaluations to identify and study typical delays in the underground train operation. As a result, we were able to create a reference timetable for our optimization which more closely matches the actual underground traffic in the system. It can be used by VAG during schedule creation to improve both the reliability of future underground timetables and the reliability of the projected energy savings by our optimization procedures in practice. The next step in this ongoing project will be to refine our timetabling model further in order to integrate some more operational requirements. At the end of this process, VAG is going to adopt the timetable planning software we are implementing based on our mathematical approaches to support planners in creating energy-efficient underground timetables.

Industrial relevance and summary

The results we obtained exhibit a significant potential for the reduction of energy consumption, and thus electricity costs, by optimized timetabling. We

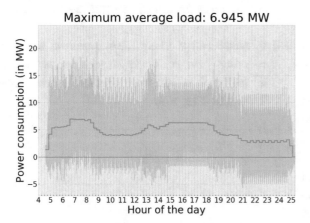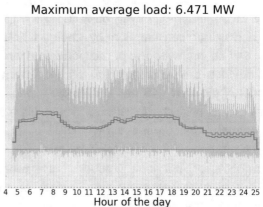

Maximum average load: 6.945 MW   **Maximum average load: 6.471 MW**

(a) The power consumption pattern induced by the Nürnberg underground timetable before (left) and after (right) optimization. The *blue* curve shows the total power drawn by all trains in the network in a given second of the day, while the *red* curve shows the average power consumption over 15-minute intervals for the unoptimized timetable. The *green* curve shows the same for the optimized timetable. Energy consumption (as the area under the *red/green* curve) can be lowered throughout the whole day, with peak consumption in the morning rush hour decreasing by almost 7%.

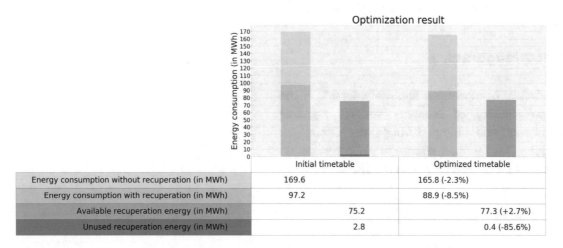

Optimization result

	Initial timetable	Optimized timetable
Energy consumption without recuperation (in MWh)	169.6	165.8 (-2.3%)
Energy consumption with recuperation (in MWh)	97.2	88.9 (-8.5%)
Available recuperation energy (in MWh)	75.2	77.3 (+2.7%)
Unused recuperation energy (in MWh)	2.8	0.4 (-85.6%)

(b) Comparison between the initial (unoptimized) timetable with the optimized timetable for some statistics related to energy consumption. Overall, the optimized timetable saves 8.5% in energy. These savings stem from both a reduced net energy consumption from the external power supply system and from an increase in the use of recuperated braking energy.

Figure 4: The potential energy savings by our optimized underground timetable

show a comparison of the unoptimized timetable draft with the optimized timetable in Figure 4. Especially, Figure 4a shows the consumption pattern over time induced by the underground timetable of the Nürnberg underground system before and after optimization. We see at first sight that the optimization leads to a sizeable decline in wasted recuperation energy as there is much less overall "negative consumption". It also turns out that the power consumption averages over consecutive 15-minute intervals, an important cost factor besides total consumption, can be lowered throughout the whole day, with peak consumption declining by almost 7%. Figure 4b shows that – according to the solution found by our optimization model – the total energy consumption can even be reduced by 8.5% through less net consumption by energy-efficient driving and an improved use of recuperation energy. These savings are twice as high as the internal threshold of 4% that VAG has set for the economic viability of this project. If realized, they would lead to an annual reduction in electricity costs of about $500,000$ € per year under optimal conditions.

This outlook of not only reducing operational costs but also the environmental impact of the underground operations has motivated us to continue our joint project in order to assess the actual savings potential when implementing the approach in practice. As already mentioned, the aim is to provide planning experts with a decision support software which automatically produces proposals for energy-optimized timetable drafts.

Acknowledgements

We gratefully acknowledge financial support by the Bavarian Ministry of Economic Affairs, Regional Development and Energy through the Center for Analytics – Data – Applications (ADA-Center) within the framework of "BAYERN DIGITAL II".

References

[1] A. Bärmann, T. Gellermann, M. Merkert, and O. Schneider. Staircase compatibility and its applications in scheduling and piecewise linearization. *Discrete Optimization*, 29:111–132, 2018.

[2] A. Bärmann, P. Gemander, and M. Merkert. The clique problem with multiple-choice constraints under a cycle-free dependency graph. *Discrete Applied Mathematics*, 283:59–77, 2020.

[3] A. Bärmann, A. Martin, and O. Schneider. A comparison of performance metrics for balancing the power consumption of trains in a railway network by slight timetable adaptation. *Public Transport*, 9(1-2):95–113, 2017.

[4] F. Liers and M. Merkert. Structural investigation of piecewise linearized network flow problems. *SIAM Journal on Optimization*, 26(4):2863–2886, 2016.

[5] B. Sansó and P. Girard. Instantaneous power peak reduction and train scheduling desynchronization in subway systems. *Transportation science*, 31(4):312–323, 1997.

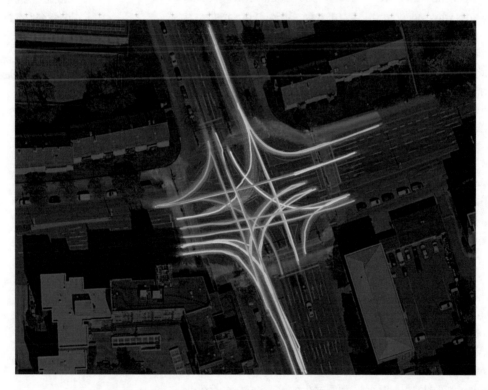

Trajectories of passenger cars, vans and trucks have been classified into their lanes (different colors) at an intersection in Lower Saxony, Germany. These data were used to learn a model of human-driven vehicles.

MATHEMATICAL OPTIMIZATION AND MACHINE LEARNING FOR EFFICIENT URBAN TRAFFIC

Traffic jams cause economical damage, which has been estimated between 10 and 100 billion Euros per year in Germany [7], also due to inefficient urban traffic [6]. It is currently open how the situation will change with upcoming technological advances in autonomous and electric mobility. On the one hand, autonomous cars may lead to an increased number of vehicles on the road with implied consequences. On the other hand, the availability of Vehicle2X (V2X) communication and smart algorithms might make the traffic flow more efficient, especially at natural bottlenecks such as urban traffic-light-controlled intersections.

To be able to quantify this anticipated potential to reduce waiting time, energy consumption, and CO2 emissions, we developed mathematical models and tailored optimization algorithms. Mathematically optimal solutions provide bounds on what could be achieved. This versatile tool can be used to analyze a large variety of scenarios, including infrastructure investments, changes of traffic-light legislation, or the interplay between humans and autonomous vehicles. Numerical results indicate that the performance indicators travel time, energy consumption and emissions could be concurrently reduced by almost 50%. Potentially, the same models and algorithms might be the basis for future traffic control systems.

To calculate optimal switching of traffic lights and optimal autonomous driving of participants, we have been developing a mixed-integer optimization model and a variety of techniques that allow an efficient computation. Scenarios include fullyautonomous as well as mixed traffic, which leads to the additional challenge of incorporating realistic and uncertain human driving behavior into the model. To this end, we have been combining methods from different areas such as discrete and continuous mathematical optimization, control theory, and machine learning. Parts of the derived algorithms were successfully implemented and tested in a car of our industrial project partner Volkswagen Aktiengesellschaft.

JOHANNA BETHGE
ROLF FINDEISEN
DO DUC LE
MAXIMILIAN MERKERT
SEBASTIAN SAGER
ANTON SAVCHENKO
Otto von Guericke
University Magdeburg

HANNES REWALD
Volkswagen Aktiengesellschaft,
Otto von Guericke University Magdeburg

PARTNERS

HANNES REWALD, Volkswagen Aktiengesellschaft, Otto von Guericke University Magdeburg
STEPHAN SORGATZ, Volkswagen Aktiengesellschaft Group Strategy – Sustainable Mobility

Industrial challenge and motivation

Urban traffic intersections are bottlenecks for an efficient traffic flow and it is an economically and environmentally important question, how to improve the performance. There are two obvious possibilities to tackle the challenge: First, to optimize the traffic light systems, and second to improve the behavior of the autonomous vehicles and human operated vehicles. Both degrees of freedom can be combined into a single, centralized optimization problem. This allows to estimate the potential for improvements of selected measures, such as introduction or modification of traffic light regulations, or the usage of optimized algorithms for autonomous driving. In addition, the approach allows to determine the capacity of a road network under maximum possible coordination, which is not only of interest for infrastructure planning purposes, but also serves as a benchmark for decentralized and other heuristic approaches.   In order to obtain valid bounds, however, one needs to solve the resulting models to global optimality. This is very challenging even on a very simple network and short time horizons. Especially, because only a few and slow approaches for globally optimal moving horizon mixed integer problems exists.

From a modeling point of view, several challenges arise. The idealized central optimization model has to be extended in several ways to cope with different types of traffic that we expect in the near future. We address the following issues for a comprehensive investigation of efficient future traffic at urban intersections:

- A centralized optimization approach that gives an idea of how a fully automated and centrally coordinated traffic would look like, and what could be theoretically achieved regarding the objective of traffic flow optimization. Given the complexity of the optimization problem, mathematical models and optimization algorithms need to be developed.

- Mixed-models that include human drivers, bicycles, and pedestrians. Such submodels must be based on insight extracted from data or from physical insight. Subsequently,

Figure 1: Trajectories of passenger cars, vans and trucks have been classified into their lanes (different colors) at an intersection in Lower Saxony, Germany. These data were used to learn a model of human-driven vehicles.

these subproblems need to be included in the centralized optimization problem formulation with new algorithms (e.g. nonlinear model predictive control (NMPC)).

- An investigation should clarify which effects can already be obtained with technology available and without huge investments. For example, driver assistance systems might facilitate some of the ideas from central optimization simply based on the knowledge of the traffic light switching. Such a system could be realized with V2X communication and NMPC, and may or may not incorporate heuristic central optimization of traffic lights.

Mathematical research

Centralized optimization

The goal of the central optimization approach is to determine the influence of optimizing certain degrees of freedom on the resulting traffic in terms of efficiency of the traffic flow (waiting times, travel times), but also on environmental quantities like fuel consumption and CO_2 emissions. The first step is to develop a suitable model to describe the movement of cars through the network and their interaction with other cars and the traffic lights. The basic structure of our scenario is as follows: Straight roads intersect in a single intersection. Each road consists of two lanes running in opposite directions. On each incoming lane, a traffic light is installed in front of the intersection that regulates the traffic on its lane. We use a simple double integrator as the longitudinal model to describe the vehicle dynamics $x = (s, v, a)$ consisting of position, velocity and acceleration. The biggest challenge in modeling and optimizing traffic at intersections is preventing collisions between cars. There are two types of possible collision situations we have to address in our modeling: First, we have to make sure that vehicles driving on the same lane do not collide, and second, we have to prevent collisions inside the intersection area, where vehicles from different lanes have to coordinate. Collision prevention constraints outside the intersection area are of the form

$$s_{c',t} - s_{c,t} \geq l_{c'} + g_c \qquad (1)$$

ensuring a sufficient distance g_c between a vehicle c and its predecessor c'. Inside the intersection area traffic lights coordinate the traffic in such a way that vehicles can safely cross the intersection. Traffic lights are modeled via binary variables, which can take the binary-encoded states red and green. The idea is that if the intersection area is occupied by a vehicle, the traffic light should give red to other lanes that can potentially cause a collision, essentially blocking the intersection for the corresponding vehicles. Indicator formulations track if a vehicle is inside the intersection area and enable or disable such constraints based on the location of the car. Special care is needed to adapt (1) for modeling turning vehicles, because the predecessor-successor relations may change after a turning maneuver.

Our main goal is to optimize traffic flow, which could be interpreted as reducing the overall travel time of all vehicles for reaching their destination. As the time horizon is fixed, minimizing travel time is similar to maximizing the covered distance at the end of the time horizon in our scenario. Therefore, we maximize the sum of the driven distances of all cars

at the last time step N:

$$\max \sum_{c \in C} s_{c,N}$$

This optimization problem is formulated as a mixed integer linear program (MILP) and yields a solution which is optimal in terms of traffic flow, but not necessarily in terms of fuel efficiency or CO_2 reduction. To achieve this objective, we follow a two-step approach. First, we optimize our model in terms of traffic flow, then we solve the same model with a changed objective function while demanding the same objctive values as obtained from the first optimization via an additional constraint and fixing all binary variables. That way, only a quadratic program (QP) has to be solved in the second step.

Table 1 shows averaged results over 5 example instances with relatively high traffic density. Centrally optimized traffic leads to almost no waiting time in case of completely free traffic lights. However, also the more realistic case of controllable traffic lights that are bound to certain regulations on phase lengths leads to a significant reduction, which is also accompanied by reductions in fuel consumption and CO_2 emission when compared to non-optimized traffic generated by the traffic simulation tool SUMO (Simulation of Urban MObility) [3]. Further computational results can be found in [4].

Decentralized optimization of mixed traffic

Another part of investigation is concerned with the efficiency of mixed traffic, namely, a situation when human drivers act alongside autonomous vehicles. This causes a need for modeling human drivers' behavior. Such models can be based on first principle modeling, machine learning or a combination of both, e.g. combining a bicycle model $f(\cdot)$ [2] with neural networks $h(\cdot)$:

$$z(k+1) = f(z(k)) + h(\sigma(\cdot), z(k)) \qquad (2)$$

$$\sigma(\xi) = \frac{1}{1 + e^{-\theta \cdot \xi}}.$$

The parameters θ of the neural network are learned during the training phase using the features ξ. The future state $z(k+1)$ (vehicle position and velocity) of the human driven vehicle is then predicted for the

density	T	dt	l		waiting time	travel time	fuel	CO_2
[Cars/min]	[s]	[s]	[s]		[s]	[s]	[l/100km]	[g/km]
20.95	60	0.5	10					
real-world simulation					14.67	57.89	10.67	248.29
fixed traffic lights					8.14	39.20	9.19	213.80
regulated traffic lights					7.39	37.86	8.96	208.47
free traffic lights					0.14	26.84	6.41	149.23

Table 1: Computational results for traffic with different degrees of optimization. All measurements are per vehicle and are geometric means over five randomized test instances. From [4, Table 4].

next time steps using the neural network. By solving the optimal control problem for an autonomous car (3) (decentral optimization and non-connected vehicles), it crosses the intersection as soon and fast as possible, while keeping a minimum distance from a human driven vehicle at all times.

$$\min_{\hat{u}(\cdot),\,\hat{s}(\cdot)} J(\hat{u}(\cdot),\,\hat{x}(\cdot))$$

$$s.t. \quad J(\cdot) = \sum_{l=k}^{k+N-1} \left(\| \hat{u}(l) \|_{R_1}^2 + \| \Delta\hat{u}(l) \|_{R_2}^2 \right)$$

$$+ \sum_{l=k+1}^{k+N} \left(\| \hat{x}(l) - x_{\text{ref}}(s) \|_{Q_1}^2 + \| \hat{s}(l) \|_{Q_2}^2 \right)$$

$$\hat{x}(k+i) = f(\hat{x}(\cdot),\,u(\cdot)), \quad \hat{x}(k) = x(k) \tag{3a}$$

$$\hat{\mathbf{z}}_{j,m}(k+i) = h(\theta_m, \mathbf{z}_j(\cdot), \hat{\mathbf{z}}_{j,m}(\cdot)), \tag{3b}$$

$$d(\hat{x}(k+i), z_{j,m}(k+i)) \geq \epsilon_d, \quad \forall i \in [0, N], \tag{3c}$$

$$\Delta u = \hat{u}(k) - \hat{u}(k-1), \quad \Delta u(k) = 0 \tag{3d}$$

The geometric path $x_{\text{ref}}(s)$ of the autonomous car is a priori defined based on the layout of the intersection, while the position along the path $\hat{s}$ and the acceleration $\hat{u}$ of the autonomous car are determined online by solving the optimal control problem (3). The path position $\hat{s}$ and the optimal acceleration profile $[\hat{u}(k), \hat{u}(k+1), \ldots, \hat{u}(k+N)]$ are constrained by the predicted position of the human driven vehicle (3b) and the minimum distance ϵ_d (3c) between the autonomous car $\hat{x}(k+i)$ and the human driven vehicle $\hat{z}(k+i)$ predicted through the neural network (2). The state $\hat{x}(k+i)$ of the autonomous car at time step $k+i$ is predicted and optimized based on the optimal acceleration profile $\hat{u}(\cdot)$ and the car model (3a). By setting $R_2 > 0$ one can incorporate the notion of travel comfort for

the occupants of the autonomous car. It aims to reduce the changes in acceleration profile (3d) of the autonomous car, which should also reduce fuel consumption. However, this is not necessary with respect to safety.

The decentralized approach was implemented for a single autonomous vehicle, which does not depend on wireless communication, as e.g. V2X or V2V. Preliminary results show that based on the uncertainty of the human driven vehicle model and the tuning parameters (R_1, R_2, Q_1, Q_2, ϵ_d) the autonomous vehicle crosses the intersection in less time and with smaller distance between the vehicles than human drivers.

We validated the decentralized optimal controller using first principle models and a neural network to model the human driving behavior (2) in [1]. A sketch of the controller's behavior is shown in Figure 2. When approaching an intersection, the autonomous car collects data on the human driven car. The autonomous car predicts the trajectory of the human driven vehicle (red arrows) and its probability (thickness of the red arrows). Based on these predictions, the acceleration (thickness of yellow arrows) of the autonomous car is optimized and results in the planned trajectory (yellow arrow) of the autonomous vehicle. Far away from the intersection the possibilities or modes (turning left or right or going straight) of the human driven car are equally likely and result in multiple predicted trajectories—one for each possibility. Due to updated information the autonomous car removes less likely possibilities and adapts its own acceleration until it safely turns at the intersection behind the human driven car. In the simulation example on the right

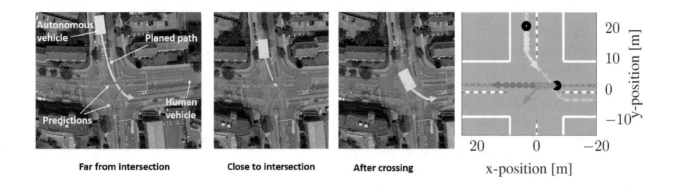

Far from intersection   Close to intersection   After crossing

Figure 2: Intersecting trajectories of the autonomous vehicle (yellow) and the human driven vehicle (red and green, without probability of each mode). A sketch of the decentral optimization concept over time can be seen on the left side, while the right side shows a corresponding simulation example based on real data.

side of Figure 2 the path of the autonomous vehicle is depicted by a yellow line. The simulation shows the planned (yellow dots) and predicted paths (red and green) at time step $k = 50$, i.e. $t = 5$ s. The future position of the human driven vehicle is predicted for two possible modes. These are going straight from east to west (red dots with direction arrow) or turning left from east to south (green crosses with direction arrow). Both predicted trajectories of the human driven vehicle are considered as constraints in the optimal control problem (3), i.e. the minimum distance $\epsilon_d = 5$ m should be kept at all time steps $k+i$ during the prediction horizon $i = [0, N]$. This leads to an optimal control input $\hat{u}(\cdot)$ for the autonomous car and the predicted position trajectory (yellow dots) at $k = 50$, shown on the right side of Figure 2. It can be seen that the minimum distance is kept for all predicted time steps at $k = 50$. Furthermore, the final trajectories for the autonomous car (yellow crosses) and the human driven car (red crosses) are shown as thin lines.

Decentralized optimization with V2X communication

Finally, a third approach realizes a combination of decentralized optimization of the velocity of each car together with a centralized control of traffic light states. Cars register with the system when

approaching a traffic light, which then greedily assigns transition trajectories on a first-come-first-serve basis with respect to some global objective. This is done after solving a simple mixed-integer program where the trajectories of all previously registered cars are fixed. If no feasible solution can be found, one or multiple correction steps are triggered which allow reassessment of the trajectories of other cars. We evaluated this heuristic method employing central optimization as a benchmark. Despite its computational simplicity, the decentralized optimization approach allowed us to reach near-optimal traffic efficiency in many instances, in particular for low to medium traffic densities.

As a preliminary step for practical testing, we considered the problem of optimally approaching a traffic light knowing its switching times, which can be realized by a V2X-based velocity controller.

Implementation

The Research Group of Volkswagen Aktiengesellschaft brought up the questions and shared their experience gained in former related projects. They also provided real-world data from cars and traffic infrastructure, and built a prototypical car, which allowed to implement and test one of the developed algorithms in real-world traffic. Parts of the real-world data were gathered from cameras on a research junction in Brunswick (see title picture). This junction is a 4-way traffic-light controlled intersection. The data set includes the traffic light signals and all traffic participants, which were categorized to cars, trucks, vans, motorcycles, bicycles, pedestrians, and railway vehicles. The data set consists of over 140.000 trajectories of motorized vehicles and includes position trajectories with time stamps and vehicle dimension separately for each car, while the velocity and acceleration trajectory were interpolated using the position and time stamps. The researchers at the Otto von Guericke University Magdeburg took the lead in the design and implementation of models and solving procedures. Especially their background in control theory—also in an automotive setting—was very fruitful for the project. The shared work of the partners was embedded in several PhD-theses [5] and in the Research Training Group GK 2297 on mathematical complexity reduction.

We considered several related approaches, which were all based on mathematical optimization. First, a velocity controller was implemented, which aims for the green phase of a traffic light. Key technology for this is the wireless communication via V2X between the traffic light and an approaching car. Multiple times per second, data about the current state of the traffic lights as well as upcoming phase switches are exchanged. This allows the calculation of velocity trajectories in the car with the aim of passing the stopping line as soon as the light switches to green and as close as possible to the driver's preferred velocity. If possible, full stops are avoided. Figure 3 gives an example illustration of such a system that has been implemented based on work from our project by Volkswagen Aktiengesellschaft. Following this, the other approaches described above were developed.

Industrial relevance and summary

The majority of research and engineering efforts concerning connected autonomous vehicles (CAV) focuses on safety matters. Talking about sustainable mobility and mobility services which aim to use as

Figure 3: Driver's view when approaching a traffic light. The driver assistance system ensures an automated passage after the light switches to green without having to stop.

few (public) resources as possible (e.g. space, time, air, and noise pollution), an efficient and smooth traffic flow needs to be achieved. Mathematical studies have shown that even single CAVs can have a positive impact on overall traffic flow [8]. Thus, research on optimizing the individual and collective driving behavior of CAVs is of current interest. In fact, human drivers are already able to adapt to different traffic situations and to implicitly perform group maneuvers. A first step of development needs to make CAVs as efficient as current human driven vehicles. Further potential beyond this should be unlocked in future development of CAVs, for which this project carves a path.

One major result was the evaluation of the effects on traffic flow for all three approaches. Best results were naturally achieved by the central optimization model. A reduction in the mean waiting time of all cars of up to 99% could be achieved. Also, the average fuel consumption could be reduced by 39%. Furthermore, the model allows to compare the influence of different parameters, such as traffic density, traffic light scheme, share of automated vehicles or speed limit, on the traffic efficiency. Surprisingly, the hybrid method achieved similar outcomes although necessary solving times are much shorter and thus offer a practical implementation. Finally, the V2X-based velocity controller led to 28% less waiting times and 19% less fuel consumption. Furthermore, it was implemented and successfully tested in a real car. Test runs in real-world traffic in the cities of Brunswick and Düsseldorf confirmed the practicability of the assistance system.

Patents

» Florian Kranke, Holger Poppe, Stephan Sorgatz: DE102015221815B3

References

[1] J. Bethge, B. Morabito, H. Rewald, A. Ahsan, S. Sorgatz, and R. Findeisen. Modelling human driving behavior for constrained model predictive control of mixed traffic at intersections. *IFAC World Congress Berlin*, pages 14557 – 14563, 2020.

[2] J. Kong, M. Pfeiffer, G. Schildbach, and F. Borrelli. Kinematic and dynamic vehicle models for autonomous driving control design. *2015 IEEE Intelligent Vehicles Symposium (IV)*, pages 1094–1099, 2015.

[3] D. Krajzewicz, J. Erdmann, M. Behrisch, and L. Bieker. Recent Development and Applications of SUMO-Simulation of Urban MObility. *International Journal On Advances in Systems and Measurements*, 5(3 and 4):128–138, 2012.

[4] D. D. Le, M. Merkert, S. Sorgatz, M. Hahn, and S. Sager. Autonomous traffic at intersections: an optimization-based analysis of possible time, energy, and CO_2 savings. Optimization Online, 2020.

[5] S. Sorgatz. *Optimization of Vehicular Traffic at Traffic-Light Controlled Intersections*. PhD thesis, Otto-von-Guericke-Universität Magdeburg, 2016.

[6] Wirtschaftswoche. Autofahrer verlieren sehr viel Zeit im Stau. `https://www.wiwo.de/politik/deutschland/154-stunden-pro-jahr-autofahrer-verlieren-sehr-viel-zeit-im-stau/23975904.html`, February 12 2019.

[7] Wirtschaftswoche. Der Stillstand kostet Milliarden. `https://www.wiwo.de/politik/deutschland/staukosten-der-stillstand-kostet-milliarden/23977168.html`, February 12 2019.

[8] Y. Zheng, J. Wang, and K. Li. Smoothing traffic flow via control of autonomous vehicles. arXiv:1812.09544, 2018.

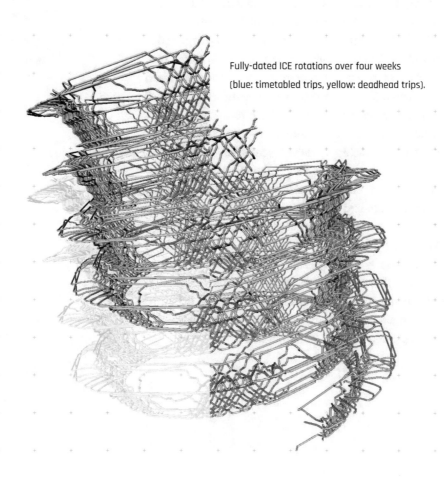

Fully-dated ICE rotations over four weeks
(blue: timetabled trips, yellow: deadhead trips).

© Springer Nature Switzerland AG 2021
H. G. Bock et al. (eds.), *German Success Stories in Industrial Mathematics*,
Mathematics in Industry 35, https://doi.org/10.1007/978-3-030-81455-7_20

TRAIN ROTATION OPTIMIZATION

Scheduling ICE High Speed Trains

Train rotation planning, i.e., the construction of efficient circulations of rolling stock in order to implement a timetable, is a key problem for railway companies. A major challenge is the formation of trains from several individual units of rolling stock, which can differ by type, position, and orientation. This feature is unique and distinguishes rail bound traffic from other modes of transportation such as air traffic or bus transit. It also prevented for several decades the use of classical network flow based vehicle scheduling methods that are well established at airlines and bus operators. This long standing barrier was breached by the development of a new hypergraphical approach, that allows to deal with train composition and schedule regularity in one common, conceptually simple modelling framework. Exploiting the special structure of train rotation problems was the key to the solution of large-scale problems by a novel coarse-to-fine method that works on aggregate model versions where possible and only resorts to more detailed variants when needed. With these innovations, it becomes possible to optimize planning scenarios with up to 100 million degrees of freedom, such as the "standard week" for an entire fleet of ICE trains in long-term planning or "rolled out" four to six-week "fully-dated" scenarios in less-than-a-year construction site planning. An optimizer RotOR has been implemented along these lines and integrated into Deutsche Bahn's FEO train rotation optimization system; it is in productive use since 2013. Estimated revenue increases amount to 24 million € per annum in Deutsche Bahn's ICE operations.

RALF BORNDÖRFER
Zuse Institute Berlin &
Freie Universität Berlin

MARKUS REUTHER
STEFFEN WEIDER
LBW Optimization GmbH

PETER SCHÜTZ
Deutsche Bahn AG

KERSTIN WAAS
DB Netz AG

Industrial challenge and motivation

Deutsche Bahn is Europe's second largest operator in long distance passenger transport. The long-distance division DB Fernverkehr (DBFV) operates 1 476 timetabled trips to transport 410 000 travelers per day, 25% more than the total number of passengers transported by Lufthansa all over the world. The turnover is 4.7 billion € per year with an EBIT of 417 million € per year [13]. The workhorses of DBFV's operations are 274 ICE high speed railcars of 10+ types, from the ICE 1, which is in operation since 1991, to the newest ICE 4 units. At a cost of up to 33 million € per piece (ICE 3 type 407/Velaro, see Figure 1), the efficient use of this rolling stock is critical. A key problem is train rotation scheduling, i.e., the construction of efficient circulations of railcars in order to implement a given timetable [5, 6].

The main challenge in train rotation planning is to handle the formation of trains from several units of rolling stock, namely, double tractions (ICE $2, 3, T$). Within a train, the units can differ by type, position (first, second), and orientation (tick = 1st class in front, tack = 2nd class in front), such that a large combinatorial variety of possible train formations results, see Figure 3 for an example. Failing to deal with train formations results in chaos at the platform. A second challenge is regularity, i.e., similarity of operations on successive days of the week. Regularity produces routine in operations and is highly desired. In particular, regular timetables are supposed to be operated by regular train rotations. Finally, the scheduling of maintenance stops is critical, as there is only one facility for every type of ICEs in Germany (e.g., the ICE 2 is maintained at the depot of Rummelsburg in Berlin). When for instance the 8 000 km inspection is due, the unit should better be already close to the depot instead of at the other end of the country. Mileage reduction by avoiding unproductive deadhead trips is clearly a major objective, second only to fleet minimization.

Mathematical research

Train rotation planning has traditionally been done with limited technical support by experienced

Figure 1: ICE 3 Velaro high speed train.

planners in an evolutionary approach, in which adjustments in the timetable were implemented by adjustments in train rotations. This worked well until the liberalization of the European railway market which started at the turn of the millennium. Since then, increasing competition on the limited railway infrastructure led to more and more disruptive changes of services that at some point called for the ability to routinely perform green field re-computations of the entire system of rotations on short notice. This was simply no longer possible in the old way. At this point, in 2008, DB Fernverkehr launched the FEO project (Fahr- und Einsatzoptimierung) to develop a decision support system for train rotation optimization. Its core is a mathematical train rotation optimizer RotOR that was developed at the Zuse Institute Berlin.

The foundation for the solution of the train rotation optimization problem is a model in terms of a hypergraph [10, 11]. This model represents the simultaneous movement of several units of rolling stock in a train by a hyperarc, see Figure 4. Different possible formations result in a set of possible hyperarcs, which can be enumerated a priori. In fact, the possible formations are governed by a set of elaborate rules that reflect technical, operational, and marketing considerations. The same holds for transitions between formations at turns, deadhead trips, joins, splits, etc. And, even better, the regularity of operations can also be expressed by forming hyperarcs over several days of operations,

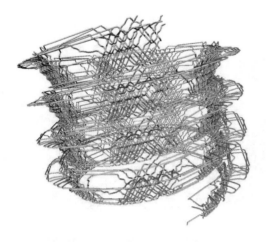

Figure 2: Fully-dated ICE rotations over four weeks (blue: timetabled trips, yellow: deadhead trips).

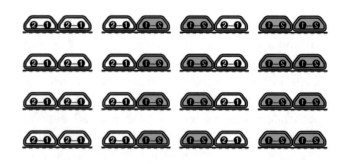

Figure 3: There are 16 possibly double tractions of two railcars of two possible types on two positions in two possible orientations; here, white=tick, gray=tack, the front is to the right.

see Figure 5 for an illustration. This means that train composition and regularity can both be handled in terms of a unified, structurally simple approach. Given the hyperarcs, all that has to be done is to enforce flow conservation of railcars at each station, i.e., if a railcar enters a station via some hyperarc, it must also leave the station via some other hyperarc. This produces a hyperflow model, which is structurally similar to a standard network flow model, except that it uses hyperarcs instead of arcs:

$$c^{\mathsf{T}} x \to \min$$
$$Ax = 0,\ Tx = 1,\ x \text{ binary}$$

Here, A is the node-hyperarc incidence matrix of the hypergraph and T is the timetabled trip-hyperarc incidence matrix. Hypergraphs enjoy at present limited popularity in computational optimization, because they typically lead to very large models with little structure that are hard to solve. One notable exception is set partitioning, which is the dominating model in all types of crew scheduling applications; its success relies on the efficient computation of small substructures. The key observation here is that a similar, albeit different, structure can be exploited in train rotation optimization. The first point is that the number of hyperarcs is of course very large, however, the "local complexity" is bounded. In fact, only a certain number of possible train formations can be run on any trip or follow one another. These

options can be produced on the fly efficiently using standard column generation techniques (like in set partitioning). But this alone is not enough. The second point is that often details of the composition such as the position or orientation of a unit of rolling stock will automatically turn out right "automatically" even without consideration. For example, if a train enters a dead-head, there is no way to turn it around, such that the positions and orientations of the unit are automatically reversed; there is just one option. If the train enters with a legal formation, it will automatically leave with a legal formation. In other cases, there is enough time to turn, e.g., in an overnight stay, such that there is also no problem. If at certain critical points the right decisions are taken, the rest of the plan falls into place. Of course, the problem is that it is not know in advance where these points are. However, these considerations inspire a procedure to set up hypergraph models of varying degrees of freedom, ranging from a very simple flow model that just tracks the movements of individual vehicles without explicitly considering train formations at all (we call this the vehicle layer), via an intermediate level which considers trains and vehicles types (the configuration layer), to a final layer of full detail (the composition layer), see Figure 6. The coarse layers lead to much smaller models than their finer counterparts, and can be merged into one model of locally different depth of detail. Each instantiation of such a model is a relaxation of the fully-detailed master model. Solving this relaxation, one hopes that all requirements come out in the right way. If this is the case, an optimal solution has been found. If not, some illegal operations are performed, e.g., reversing where turning is not allowed. Then the model is locally refined. At other places, we might conclude that a refinement is no longer needed and coarsen. If this coarse-to-fine method [8] is implemented in the right way, it converges to the global optimum (or in practice to a solution with a small duality gap) [9]. This is the essence of the RotOR optimizer.

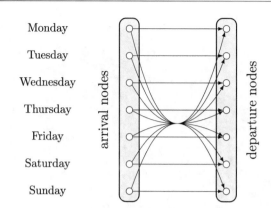

Figure 5: The regular operation of an everyday tip on all days of the week can be modeled using a regularity hyperarc.

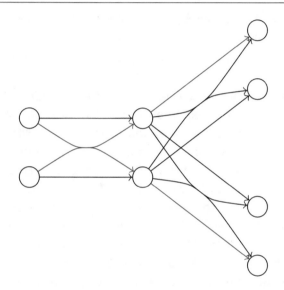

Figure 4: Idea of the hyperflow train rotation model: the simultaneous operation of two vehicles in a double traction train to service a trip can be modeled in terms of a hyperarc (the blue hyperarc on the left). After its arrival, the double traction is split and the two individual units service different trips in single tractions (represented by the blue arcs). The black arcs and hyperarcs represent alternative options.

The coarse-to-fine method exploits the special structure of hypergraphs arising from train rotation optimization problems, namely, that each hyperarc represents a union of standard arcs. We call such hypergraphs graph-based. We believe that the identification of such structures can give rise to an algorithmic theory of hypergraphs similar to algorithmic graph theory. A number of theoretical results for related classes of hypergraphs, including a generalized Hall theorem for normal hypergraphs [3], a tight cut decomposition algorithm for

matching-covered uniformizable hypergraphs [2], and a purely combinatorial hypergraph network simplex algorithm [1] corroborate this hypothesis.

We remark that train rotation optimization must meet a number of demands in addition to train formation, in particular, maintenance requirements. These are handled by an additional model component in terms of "resource flows". The complete model is discussed in a number of articles [10, 8, 9] and extensively in the PhD thesis of Reuther [14].

Implementation

The development of RotOR started in 2008 as a cooperation between DB Fernverkehr and ZIB. The project went through several phases, until it passed over into the RailLab of the BMBF Research Campus MODAL, in which a fully-dated version of the optimizer for the solution of acyclic problems with boundary constraints was developed. Investigations on the structure of hyperassignment and hyperflow problems were additionally carried out in the second phase of the BMBF Research Center Matheon – Mathematics for Key Technologies (Application Area Networks, Project B22).

RotOR 1.0 was integrated into DB Fernverkehr's FEO system, see Figure 7 left, and first went into production in the strategic planning department in June 2013. Release 2.0 of March 2014 reduced memory consumption and implemented a full fledged adaptation optimization. RotOR 3.0 of

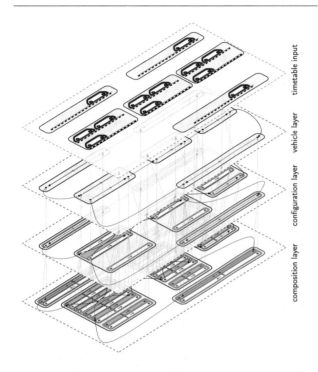

Figure 7: Screenshot of DB Fernverkehr's FEO train rotation planning system.

Figure 6: The three layers of the coarse-to-fine model are compositions, configurations, and vehicle flows.

May 2018 could handle dynamic time horizons; the implementation of this feature also included a thorough refactoring of the entire code base. The current version RotOR 4.0 allows the solution of fully-dated, rolled out scenarios (see Figure 2) to deal with less-than-a-year-away planning of construction sites. Maintenance is provided by the ZIB spin-off company LBW Optimization GmbH.

Industrial relevance and summary

DB Fernverkehr first used FEO/RotOR in the strategic planning department for supporting the dimensioning of its ICE 4 fleet. Extensive scenario analyses were performed in order to determine how many railcars of which sizes (numbers of coaches) should be procured and whether the units should feature clutches to allow for double tractions (like the ICE 2) or not (like the ICE 1). However, the main application scenario at DB Fernverkehr is construction site management. DB Fernverkehr is struggling with a record-high of currently 800 construction sites per day, such that it becomes increasingly difficult to pass through bottlenecks and to bypass temporarily closed track segments. These construction sites result from a 156 billion € investment program of the German government and Deutsche Bahn to meet the increasing demand for rail transport services (long distance passenger transport has increased by almost 50% since 1995 [12]. For the time being, however, the construction sites are a problem. They are tackled using adaptation optimization, i.e., the adaptation of some existing (mostly a former) solution to exogenous or endogenous changes, e.g., changes in trip times because of improvements or deteriorations of track segments, construction sites etc. The input "reference solution" will typically be partially infeasible and has to "repaired", and the fix is usually sought as a compromise between staying close to the reference solution and minimal operation costs, i.e., one can (and must, if the reference solution is infeasible) deviate, if it pays off, otherwise one shouldn't. This scenario is particularly well suited to the coarse-to-fine method and solves much faster than a green field optimization in which the optimizer has less guidance to the desired outcome [7, 4]. RotOR supports adaptation optimization by automatically configuring its objective function to reproduce as many details of the reference rotations as possible.

A major success was the management of the 2015

Cologne-Rhine-Main construction site, see Figure 8. In April and May 2015, the 117 km high-speed link between Cologne and Frankfurt was shut down on four succeeding weekends because of construction work, such that the traffic had to use the old route through the Rhine valley. The resulting driving times were incompatible with DB's synchronized timetable. The problem affected the high speed traffic of the entire ICE 1 fleet. There was no manual solution that could do with the available number of ICEs. With RotOR, a "minimally invasive" solution was found, that limited unavoidable trip cancellations to a small and operationally separable set of lines. Line 82 to Brussels could continue its operations without any changes. This was important, as the ICE vehicles on this line feature special equipment for entering neighboring countries. The weekend trips of three lines (45, 47, and 49) had to be canceled, but the remaining five lines (41, 42, 43, 78, and 79) could continue their services, complete with the necessary weekend phase-ins and outs.

Four construction sites from the fourth quarter of 2019 have been analyzed in detail in order to quantify the benefits of RotOR for less-than-a-year-away construction site management. These were the construction sites "Gelnhausen" (3 days, start of Oct), "Offenbach" (3 days, middle of Oct), "Dollbergen-Fallersleben" (6 days, start of Nov), and "Ulm" (38 days, Nov/Dec). In these scenarios, FEO prevented service cancellations of 28 200 km, 21 300 km, 8 100 km, and 304 000 km, respectively, i.e., around 360 000 train km in total. Extrapolating these numbers to a full year results in more than 1 000 000 additional train km, which is the annual mileage of 2 ICE double traction units. This in turn translates into 784, 000 additional customers and 34 000 tons of CO_2 savings per year. The resulting revenue increase for the year 2019 has been estimated by DB Fernverkehr at 24 million €.

Although most of the savings in this scenario come from one large construction site the many small sites also pose a significant problem. One cannot afford significant manual planning for each of them, and in particular not in the face of constant shifts of timings (the exact duration of construction work is difficult to predict), and the combinatorial complexity that results from the interaction of all these sites

is apparent. Integrated planning systems like FEO, including powerful optimizers such as RotOR, are needed to deal with such operational challenges, and to seize the opportunities that arise from the growing demand.

The FEO decision support system with the RotOR optimization kernel has greatly increased Deutsche Bahn's ability for scenario analysis, enhances reaction speed, and improves strategic asset allocation.

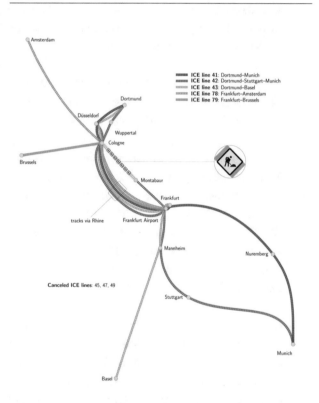

Figure 8: The ICE rotations through the 2015 Cologne-Rhine-Main construction site were managed by FEO.

Acknowledgements

The authors acknowledge the support by the Federal Ministry of Education and Research (BMBF Research Campus MODAL – Mathematical Optimization and Data Analysis Laboratories, RailLab, and BMBF Research Center MATHEON – Mathematics for Key Technologies, project B22).

References

[1] I. Beckenbach. A hypergraph network simplex algorithm. In N. Kliewer, J. F. Ehmke, and R. Borndörfer, editors, *Operations Research Proceedings 2017*, pages 309–316. Springer Verlag, 2018.

[2] I. Beckenbach. *Matchings and Flows in Hypergraphs*. PhD thesis, Freie Universität Berlin, 2019.

[3] I. Beckenbach and R. Borndörfer. Hall's and konig's theorem in graphs and hypergraphs. *Discrete Mathematics*, 341(10):2753 – 2761, 2018.

[4] R. Borndörfer, B. Grimm, M. Reuther, and T. Schlechte. Template-based re-optimization of rolling stock rotations. *Public Transport*, pages 1 – 19, 2017.

[5] R. Borndörfer, M. Grötschel, and U. Jaeger. Planungsprobleme im öffentlichen Verkehr. In M. Grötschel, K. Lucas, and V. Mehrmann, editors, *PRODUKTIONSFAKTOR MATHEMATIK – Wie Mathematik Technik und Wirtschaft bewegt*, acatech DISKUTIERT, pages 127–153. acatech – Deutsche Akademie der Technikwissenschaften und Springer, 2008. ZIB Report 08-20.

[6] R. Borndörfer, T. Klug, L. Lamorgese, C. Mannino, M. Reuther, and T. Schlechte, editors. *Handbook of Optimization in the Railway Industry*. Number 268 in International Series in Operations Research & Management Science. Springer International Publishing, 2018.

[7] R. Borndörfer, J. Mehrgardt, M. Reuther, T. Schlechte, and K. Waas. Re-Optimization of Rolling Stock Rotations. In D. Huisman, I. Louwerse, and A. P. Wagelmans, editors, *Operations Research Proceedings 2013*, pages 49 – 55. Springer Verlag, 2014. ZIB Report 13-60.

[8] R. Borndörfer, M. Reuther, and T. Schlechte. A coarse-to-fine approach to the railway rolling stock rotation problem. In *14th Workshop on Algorithmic Approaches for Transportation Modelling, Optimization, and Systems*, volume 42, pages 79 – 91, 2014. ZIB Report 14-25.

[9] R. Borndörfer, M. Reuther, T. Schlechte, K. Waas, and S. Weider. Integrated optimization of rolling stock rotations for intercity railways. *Transportation Science*, 50(3):863 – 877, 2016.

[10] R. Borndörfer, M. Reuther, T. Schlechte, and S. Weider. A Hypergraph Model for Railway Vehicle Rotation Planning. In A. Caprara and S. Kontogiannis, editors, *11th Workshop on Algorithmic Approaches for Transportation Modelling, Optimization, and Systems (ATMOS 2011)*, volume 20 of *OpenAccess Series in Informatics (OASIcs)*, pages 146–155, Dagstuhl, Germany, 2011. Schloss Dagstuhl–Leibniz-Zentrum fuer Informatik. ZIB Report 11-36.

[11] R. Borndörfer, M. Reuther, S. Thomas, and S. Weider. Vehicle Rotation Planning for Intercity Railways. In J. C. Muñoz and S. Voß, editors, *Proc. Conference on Advanced Systems for Public Transport 2012 (CASPT12)*, 2012. ZIB Report 12-11.

[12] Deutsche Bahn AG. Wettbewerbsbericht, 2004–2014.

[13] Deutsche Bahn AG. Zahlen & Fakten 2018, 2019.

[14] M. Reuther. *Mathematical Optimization of Rolling Stock Rotations*. PhD thesis, Technische Universität Berlin, 2017.

Water crossing: advanced MESHFREE simulations

MESHFREE SIMULATIONS IN CAR DESIGN: CLOSING THE GAPS OF CLASSICAL SIMULATION TOOLS

Numerical simulation and digital twins for the automotive industry

Numerical simulation speeds up the design process of products tremendously. In numerous cases, simulations cannot cover the complete spectrum of the product's functionalities and its production processes. Usually, there are modeling tasks, for which classical simulation tools are unfeasible as they require too much human and computation resources, or fail completely. Thus, new simulation ideas are required which allow to bridge those gaps

and provide true digital twins in product design.

MESHFREE is a simulation tool for fluid flow and continuum mechanics. It avoids FE-meshes and, thus, is extremely flexible. It has the potential to overcome some of the gaps that are left open by classical FE-simulation tools. In this article, we describe the success of MESHFREE in the design of new cars.

JÖRG KUHNERT
Fraunhofer ITWM, Kaiserslautern

PARTNERS

LARS ASCHENBRENNER, STEPHAN KNORR, **Volkswagen AG, Wolfsburg**

STEFFEN HAGMANN, **Porsche AG, Weissach**

DIRK BÄDER, **Audi AG, Ingolstadt** | ALAIN TRAMECON, **ESI Group, Paris**

Industrial challenge and motivation

Digital twins will be the future of product design in all aspects, as they allow virtual development and layout of design prior to assembling the first prototype. Many critical aspects of new products will be identified, solved, and optimized with the help of digital twins. Thus, construction of hardware prototypes can be avoided, saving money, material, and human resources.

The basis of a digital twin is a physical or mathematical model of the product behavior under operational conditions, of its properties, and of its production process. In many cases, these models cannot be solved in a closed manner. Thus, only numerical simulation can provide the solution.

In car production, a digital twin has to mirror the behavior of a vehicle under realistic operational conditions. All aspects of durability, performance, stability, comfort, and safety have to be modeled up to a sufficient precision in short time.

Many modeling aspects suffer from insufficient numerical methods. Let us pick three of them where mathematical research and development have brought up new, better simulation methods bridging the gaps and now serving as standard tools in car design.

Challenge 1: *Numerical models in crash safety of cars*

A car has to satisfy certain requirements on safety and crashworthiness. The stability of the car body during a crash is one of them, the secure functionality of the restraint and airbag systems is another. The deployment of airbags is an example of complex Fluid-Structure-Interaction (FSI): the gas dynamics inside of the airbag, the dynamic behavior of the airbag membrane, and the interaction of the membrane with the human body. Each process for itself is a hard modelling task, which becomes even harder in the FSI context. The gas flow inside of the airbag cannot be modeled with classical, mesh-based tools due to the rapid change of the flow domain. New simulation methods are required.

Challenge 2: *Numerical models in water management of cars*

Figure 1: Watercrossing: advanced MESHFREE simulations.

Water management is a wide topic: aquaplaning, the drain of rainwater, impact of spouting water onto the vehicle body, formation of spray and soiling effects. Prior to the numerical simulation, nothing is known about where the water will go, will it spout left or right, will it enter the engine compartment, will spray cover the mirrors or side windows. Thus, for classical mesh-based simulation tools, even for the analysis of 1-mm-droplets, the whole car and its surrounding would have to be wrapped by a simulation mesh, making the simulation unfeasible. This emphasizes the need for new approaches.

Challenge 3: *Numerical models in car interactions with granular material*

Cars driving on sand or gravel might run into critical situations. For example, the car may or may not turn over if going through a sharp bend. There has to be a virtual understanding if and why it rolls over in order to trigger airbags protecting driver and passengers. A scenario like this is extremely difficult to test in reality. However, it is also difficult to model on a numerical basis. Sand is a continuous material on the one hand suggesting a mesh-based simulation. On the other hand, under highly kinetic interaction with a vehicle's body, sand might spray out such that classical mesh-based tools are unfeasible. Hence, again, simulation alternatives are strongly requested.

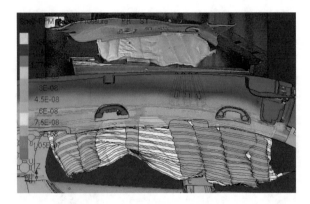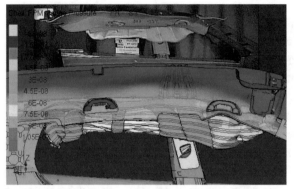

Figure 2: Airbag deployment: comparison of experiment and simulation of a curtain airbag.

Mathematical research

The mentioned shortfalls of classical mesh-based simulations arise if the flow domain changes in time, fluids contain dynamic free surfaces with droplet formation, and different phases interact with each other.

Fraunhofer ITWM started the development of the numerical tool MESHFREE in 2000. Since 2015, it is a joint cooperation with Fraunhofer SCAI. Its purpose is to model fluid flows and processes in continuum mechanics without the help of meshes. MESHFREE is based on arbitrary point clouds. The points densely cover the computational domain. Each point carries all relevant information. Since it moves with material velocity, the point cloud automatically adjusts to changes in the geometry (moving parts, deformations in FSI, etc.), as well as to free surfaces and phase boundaries. The management of the moving point cloud is simple. In fact, local refinement or coarsening does not require additional computational effort, thus adjusting the local computational accuracy and saving computation time.

The former name of the method is Finite Pointset Method (FPM), still being used in the context of airbag deployment (see further down). The approximation and modelling idea is a generalized Finite Difference Method (GFDM) [5], which provides a very flexible approximation in MESHFREE context. As the points are purely numerical, we can easily add or remove points, which gives rise to a truly adaptive

numerical scheme, see [2]. The conservation laws are modelled in their differential form:

$$d\rho/dt + \rho(\nabla^T \boldsymbol{v}) = 0 \quad \text{mass} \tag{1}$$

$$d(\rho\boldsymbol{v})/dt + \rho\boldsymbol{v}(\nabla^T \boldsymbol{v}) = -\nabla p + (\nabla^T \boldsymbol{S})^T + \boldsymbol{g} \quad \text{momentum}$$

$$d(\rho E)/dt + \rho E(\nabla^T \boldsymbol{v}) = -\nabla^T(p\boldsymbol{v}) + \nabla^T(\boldsymbol{S}\boldsymbol{v}) \quad \text{energy}$$

$$+ \rho(\boldsymbol{g}^T \boldsymbol{v}) + \nabla^T(k\nabla T)$$

This set of partial differential equations (PDE) is solved locally on each numerical point. The term d/dt denotes the change-rate of some physical quantity of a point moving with fluid velocity. The material properties are settled in the stress tensor S and in the heat conduction k, and we need to provide thermodynamic closure relations $p = p(\rho, \rho\boldsymbol{v}, \rho E)$, $T = T(\rho, \rho\boldsymbol{v}, \rho E)$. The solution of the PDE requires the computation of local derivatives, such as $\nabla^T v$ and ∇p, which is achieved by the so-called moving least squares approach. Here, around each point (i.e. some space position $\boldsymbol{y}$) and with respect to some given function u, we search for the local best-fit polynomial $a_{\boldsymbol{y}}^u$ by minimizing the functional

$$\sum_{j=1}^{N(\boldsymbol{y})} W^2(\boldsymbol{y}, \boldsymbol{x}_j) \cdot (a_{\boldsymbol{y}}^u(\boldsymbol{x}_j) - u_j)^2 \overset{!}{=} \min \tag{2}$$

$$\text{with} \quad W(\boldsymbol{y}, \boldsymbol{x}_j) = \exp(-\frac{c}{h^2}(\boldsymbol{y} - \boldsymbol{x}_j)^2)$$

with $N(\boldsymbol{y})$ the number of neighbor points around $\boldsymbol{y}$, u_j the discrete function values at the locations $\boldsymbol{x}_j$, and $w(\boldsymbol{y}, \boldsymbol{x}_j)$ the generic weight function with coefficient c and local interaction radius h to be chosen by the user. The gradient of any function can be approximated by the gradient of the best fit polynomial, for example $\nabla p(\boldsymbol{y}) \approx \nabla a_{\boldsymbol{y}}^{p}(\boldsymbol{y})$, second order derivatives like the heat diffusion can be approximated as $\nabla^{T}(k\nabla T)(\boldsymbol{y}) = k\Delta T + \nabla^{T}k \cdot \nabla T \approx k\Delta a_{\boldsymbol{y}}^{T}(\boldsymbol{y}) + \nabla^{T}a_{\boldsymbol{y}}^{k}(\boldsymbol{y}) \cdot \nabla a_{\boldsymbol{y}}^{T}(\boldsymbol{y})$.

Thus, with (2), we are able to represent any derivative needed in order to numerically integrate the set of PDE in (1).

Implementation

Challenge 1: *MESHFREE in airbag deployment*
Already in 2000, FPM was incorporated into the crash-simulation software PamCrash, today referred to as VPS (Virtual Prototyping System), see [6]. The requirement of crash modelling together with airbag deployment was immense. A crash test is massively expensive. Simulation is more efficient and cheap. Until the 1990s, the gas dynamics inside of the airbag was modeled by a one-chamber-approach. Later, all attempts to capture the flow by classical mesh-based methods failed.

Figure 3: Rain: simulation of the interaction of droplets and surrounding airflow.

Hence, FPM was a true alternative. The dynamic airbag volume and membrane movement are handled by FPM without additional effort. Thus, FPM was incorporated as a function into VPS. The membrane and the human body still is handled by VPS, the gas dynamics by FPM. The data exchange between the two methods is simple due to the grid less formulation, a true success in numerical FSI, see [3].

Challenge 2: *MESHFREE in water management*
Water management is a problem of scales. One would like to study the transient flow of droplets, or layers and films of water, see [1]. Examples are the detection of droplets entering critical zones inside of the car, driving stability in aquaplaning, and soiling effects in fog environment.

The geometrical length scale of the water might be much smaller than the car. A 0.5mm spray droplet compared to the 5m length of the car spans 4 orders of magnitude. 5mm rain droplets: 3 orders of magnitude. A 5cm deep water puddle: 2 orders. Water wading in 50cm depth: 1 order. The bigger the ratio, the more expensive is the simulation. MESHFREE is predestined to handle these situations with least additional effort.

First applications were run in water wading of cars. These simulations are the least critical. Success was achieved already in 2013. The second application was the water flow on local areas, such as the opening of a trunk lid covered by water. This scenario is characterized by 1 to 2 orders of magnitude difference regarding length scale. Success was achieved in 2015. The latest success in 2018 is the study of a full car in heavy rain. Thus, there are 2 to 3 orders of magnitude difference.

Currently, efforts are made towards 4 orders of magnitude, appearing in the spray formation induced by rolling tires.

Challenge 3: *MESHFREE in car rollover simulations*
The most recent success of MESHFREE is the rollover simulation of cars in granular material (sand or gravel). Granular material might behave like an elastic or solid body. But if the yield stress is exceeded, it starts to flow like a liquid. In this case, it can even spray out or have very dynamic free

surfaces on highly dynamic impact.

Also here, MESHFREE is the perfect candidate for numerical modeling. It provides distinct advantages compared to classical simulation ideas since it is capable to simulate car-sand-interactions in acceptable time with acceptable quality.

Figure 4: Rollover: comparison of a simulation (yellow car) and experimental data (white car) for a real prototype.

The simplest way of modeling sand is the Drucker-Prager model, ref. [4]. It correlates the local yield stress with the local hydrostatic pressure. More advanced models are Hypoplasticity and Barodesy. They are characterized by highly non-linear behavior. For the car-sand-interaction, MESHFREE works successfully with the Drucker-Prager model, allowing for standard simulation runs in new design cycles. The incorporation of Hypoplasticity as well as Barodesy is still on a scientific level.

Industrial relevance and summary

In the design process of new products such as cars, numerical simulations can save time, human resources, and a lot of material. Hardware tests are usually extremely time consuming and expensive. The computer simulation of the same thing is faster, cheaper, and provides even more insight. The complete virtualization of the product, its development and production (digital twin) is one of the biggest industrial challenges of today.

MESHFREE as a mathematical simulation tool has become a standard software in car industry and is an important part of the virtualization process in the product design and development.

References

[1] A. Jefferies, J. Kuhnert, L. Aschenbrenner, and U. Giffhorn. Finite pointset method for the simulation of a vehicle travelling through a body of water. *M. Griebel and A. M. Schweitzer, editors, Meshfree Methods for Partial Differential Equations VII*, pages 205–221, 2015.

[2] J. Kuhnert. Meshfree numerical scheme for time dependent problems in fluid and continuum mechanics. *S. Sundar, editor, Advances in PDE Modeling and Computation*, pages 119–136, 2014.

[3] J. Kuhnert, I. Michel, and R. Mack. Fluid structure interaction (fsi) in the meshfree finite pointset method (fpm): Theory and applications. *M. Griebel and A. M. Schweitzer, editors, Meshfree Methods for Partial Differential Equations IX*, 129:73–92, 2019.

[4] I. Michel, I. Bathaeian, J. Kuhnert, D. Kolymbas, C.-H. Chen, Polymerou, Vrettos, and Becker. Meshfree generalized finite difference methods in soil mechanics - part ii: numerical results. *International Journal on Geomathematics*, 2017.

[5] P. Suchde, J. Kuhnert, and S. Tiwari. On meshfree gfdm solvers for the incompressible navier-stokes equations. *Computers Fluids*, 165:1–12, 2018.

[6] A. Tramecon and J. Kuhnert. Simulation of advanced folded airbags with vps-pamcrash/fpm: Development and validation of turbulent flow numerical simulation techniques applied to curtain bag deployments. *SAE Technical Paper.*, 2013.

Electromobility: robust e-machine design through the prism of computer-aided simulations.

PASIROM: PARALLEL SIMULATION AND ROBUST OPTIMIZATION OF ELECTRO-MECHANICAL ENERGY CONVERTERS

A digital twin for optimizing electromobility

Digital twins are used in industry to understand the life cycle of engineering products. Particularly during product design, simulations are important, e.g., to optimize the geometry or to investigate possible sources of uncertainties. The mathematical model behind the twin is typically a set of partial differential equations. In the case of electrical engineering this is typically given in terms of Maxwell's equations. Their time-domain solution, e.g., (for computing) the electromagnetic fields in an induction machine, is computationally very expensive due to the need of a very fine resolution along the time axis. The situation becomes even more critical during initial design stages, when the steady-state operating characteristics of electromagnetic devices have to be considered. This requires the execution of transient calculations until the stationary regime is reached and leads to further increase of the computational time, especially when an analyzed system features a very long settling time.

Parallel-in-time methods such as the Parareal algorithm have recently gained interest as a powerful acceleration tool because of their capability to distribute the workload among multiple processing units. This success story summarizes the development of new time-parallelization approaches, their application in the simulation of an induction motor of an electric vehicle (at Robert Bosch GmbH) and the knowledge transfer to industry. As a result, exploiting 80 parallel processors, the steady-state analysis can be performed up to 28 times faster (than with a) classical sequential time-stepping approach. This significant speed up is a great aid to industry, since it significantly speeds up the design workflow.

IRYNA KULCHYTSKA-RUCHKA
SEBASTIAN SCHÖPS
Computational Electromagnetics
(CEM) Group, Technical University
of Darmstadt

MICHAEL HINZE
Mathematical Institute,
University of Koblenz-Landau

STEPHANIE FRIEDHOFF
Applied Computer Science Group,
University of Wuppertal

STEFAN ULBRICH
Department of Mathematics,
Technical University of Darmstadt

PARTNERS

OLIVER RAIN, Robert Bosch GmbH

Industrial challenge and motivation

Electro-mechanical energy converters, such as electrical motors, are an essential part of our everyday lives.   We use them in household appliances, power tools, e.g., electric drills, and nowadays for e-mobility, like in e-bikes or e-cars. In particular, the 4-pole squirrel cage induction machine [1] developed at the Robert Bosch GmbH for electric vehicles, was simulated for this success story [2].   The wide exploitation range of such induction machines is caused by their relatively cheap costs, due to the absence of expensive permanent magnets, and their still relatively high efficiency.

Prior to construction of physical prototypes, industry takes advantage of computer-aided design and engineering. Mathematical modeling, multiphysical simulation and optimization allow to find new designs of machines, which further reduce energy consumption, are more environment-friendly and have a longer lifetime.   However, high-fidelity simulations are extremely time consuming, especially when one is interested in the steady-state behavior of induction machines. This often leads to the application of less accurate models which may result in suboptimal design choices. For this reason, novel efficient numerical algorithms are of paramount importance for further progress in engineering.

Mathematical research

Time-domain parallelization approaches, such as the Parareal algorithm [9] or the multigrid reduction in time (MGRIT) [4], have a great acceleration potential due to the easy work distribution among available multiple processing units of modern computer architectures.   Parareal was originally introduced and analyzed under smoothness assumptions of the inputs.   Novel Parareal-based algorithms able to rigorously handle systems, excited with discontinuous pulse-width-modulated (PWM) signals, were recently developed by the authors in [7] and [8]. Performance of MGRIT applied to problems with PWM-inputs was illustrated in [5].

Parareal is used to solve initial value problems (IVPs)

Figure 1: Electromobility: robust e-machine design through the prism of computer-aided simulations. GetDP model of an electric machine on the foreground. Background photo credit: Bosch.

for time-dependent differential equations on the interval $(T_0, T_N]$ starting from a given initial value $\mathbf{u}_0$. The method includes splitting of the time interval into N windows $T_0 < T_1 < \cdots < T_N$, separate solution on $(T_{n-1}, T_n]$, and iterative update of the sought solution $\mathbf{U}_n$ at synchronization points $T_n, n = 1, \ldots, N$ until the jumps vanish up to a prescribed tolerance (see Figure 2). Essential ingredients of Parareal are two propagators: fine $\mathcal{F}$ and coarse $\mathcal{G}$, applied on each subinterval. We denote by $\mathcal{F}/\mathcal{G}(T_n, T_{n-1}, \mathbf{U}_{n-1})$ the fine/coarse solution of the IVP at T_n, calculated starting from the value $\mathbf{U}_{n-1}$ at T_{n-1}. The Parareal iteration reads:

$$\mathbf{U}_0^{(k)} := \mathbf{u}_0,$$
$$\mathbf{U}_n^{(k)} := \mathcal{F}(T_n, T_{n-1}, \mathbf{U}_{n-1}^{(k-1)}) + \mathcal{G}(T_n, T_{n-1}, \mathbf{U}_{n-1}^{(k)}) \quad (1)$$
$$- \mathcal{G}(T_n, T_{n-1}, \mathbf{U}_{n-1}^{(k-1)})$$

for $n = 1, \ldots, N$ and $k = 1, 2, \ldots, K$. The main idea in [7] is the application of an approximative coarse propagator $\tilde{\mathcal{G}}$ which only accounts for the low frequency components of the fast switching PWM signal.

Various approaches for faster calculation of the steady state are known from the literature.   For example, computation of an appropriate initial value for the time stepping [3] or the (simplified) time-

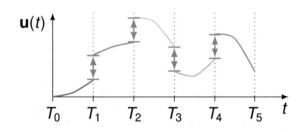

Figure 2: Splitting of the time interval within Parareal for $N = 5$. Iterative procedure is performed until the jumps of the solution at the synchronization points are reduced up to a given tolerance.

periodic explicit error correction (TP-EEC) performed after each (half) period [10] are observed to accelerate the sequential time integration. Oriented towards solution of time-periodic problems, a periodic Parareal algorithm with initial value coarse problem (PP-IC) was introduced in [6]. The difference to Parareal is only in the update of the initial value in PP-IC: $\mathbf{U}_0^{(k)} := \mathbf{U}_N^{(k-1)}$. PP-IC was

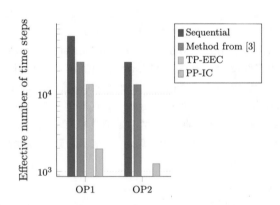

Figure 3: Computational costs of the steady state calculations for two OPs of a four-pole induction motor using different solution approaches. Simplified TP-EEC is applicable only to OP1 [2]. PP-IC applied to OP1 and OP2 parallelizes the computations across $N = 80$ and $N = 154$ processors, respectively.

recently applied to simulation of a three-phase induction motor for an electric vehicle drive by Bast et al. [2]. Periodic steady state solutions of two representative operating points (OPs) driven by

current excitation were calculated using PP-IC on the corresponding rotor periods $[0, 0.0311]$ s for OP1 and $[0, 0.114]$ s for OP2. Figure 4 illustrates the torque evolution obtained via the standard sequential time integration starting from zero initial condition and the periodic torque computed by PP-IC. Parallelization across $N = 80$ processors in the case of OP1 and $N = 154$ processors for OP2 delivered the steady state 28 and 20 times faster than classical time stepping, respectively.

Additionally to the sequential computations, performance of PP-IC is compared also to the simplified TP-EEC and to the approach presented by Bermúdez et al. [3]. The computational efforts are estimated by the effective number of calculated time steps until the steady state is reached (Figure 3). The simplified TP-EEC was applicable only to OP1 and gave the speed up of a factor of 4, while the method of Bermúdez et al. was about 2 times faster for both OPs compared to the classical sequential solution. These results demonstrate superiority of the parallel-in-time methods for the steady state analysis of induction machines.

Implementation

This work was carried out within the framework of the collaborative project PASIROM[1] (Technische Universität Darmstadt, Universität Koblenz-Landau and Bergische Universität Wuppertal) funded by the German Federal Ministry of Education and Research (grant No. 05M2018RDA) in cooperation with the industrial partner Robert Bosch GmbH and Dassault Systèmes, Deutschland.

The considered parallel-in-time algorithms were implemented in C++ using parallel processing with MPI and incorporated into Bosch's in-house finite element solver EDYSON for electromagnetics. At each iteration EDYSON was called to calculate the fine and the coarse solutions. This was possible due to the non-intrusiveness of the applied Parareal-based algorithms, which allow to exploit the existing software packages as the black-box solvers.

[1] www.pasirom.de

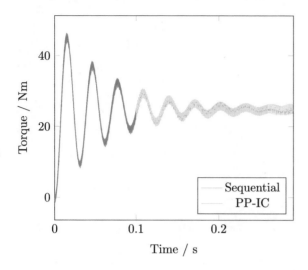
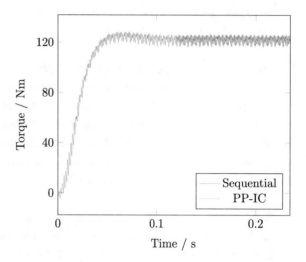

Figure 4: Sequential time stepping and the periodic PP-IC solution for OP1 (left) and OP2 (right).

Industrial relevance and summary

Due to the coordinated work of academia and industry within the PASIROM project novel efficient time parallelization approaches were developed and exploited to accelerate the numerical simulations of electromagnetic applications.

Recently, both TP-EEC and Parareal time parallelization were incorporated into the in-house finite element solver of the Robert Bosch GmbH in cooperation with the group from Koblenz-Landau and the CEM Group from Darmstadt, respectively. Effectivity of the considered parallel-in-time algorithms was illustrated in comparison to several existing alternative methods via their application to an induction motor of an electric vehicle drive, e.g., [2].

These significant advances in the time-domain calculation of electric motors obtained by means of the parallel-in-time techniques allow more freedom for simulations in the future. In particular, more complex models and even a larger numbers of different operating points can be computed within a shorter runtime for efficient and more precise treatment of the real-life demands in electrical engineering. Future research will investigate how to efficiently combine parallel-in-time methods and

robust optimization, e.g., based on [11].

Acknowledgments

The authors acknowledge the support by the Federal Ministry of Education and Research (BMBF project no. 05M2018, PASIROM).

References

[1] Alexander, M., Kelleter, A., De Larramendi, M.R., Heuser, P., Herranz Gracia, M.: Electric machine (2017). Patent US US 2017 / 0331353

[2] Bast, D., Kulchytska-Ruchka, I., Schöps, S., Rain, O.: Accelerated steady-state torque computation for induction machines using parallel-in-time algorithms. IEEE Trans. Magn. **56**(2), 1–9 (2019). DOI 10.1109/ TMAG.2019.2945510

[3] Bermúdez, A., Gómez, D., Piñeiro, M., Salgado, P.: A novel numerical method for accelerating the computation of the steady-state in induction machines. Comput. Math. Appl. **79**(2), 274–292 (2019). DOI 10.1016/j.camwa.2019.06.032

[4] Falgout, R.D., Friedhoff, S., Kolev, T.V., MacLachlan,

S.P., Schroder, J.B.: Parallel time integration with multigrid. SIAM J. Sci. Comput. **36**(6), C635–C661 (2014). DOI 10.1137/130944230

[5] Friedhoff, S., Hahne, J., Kulchytska-Ruchka, I., Schöps, S.: Exploring parallel-in-time approaches for eddy current problems. In: I. Faragó, F. Izsák, P.L. Simon (eds.) Progress in Industrial Mathematics at ECMI 2018, *The European Consortium for Mathematics in Industry*, vol. 30. Springer, Berlin (2019). DOI 10.1007/978-3-030-27550-1_47

[6] Gander, M.J., Jiang, Y.L., Song, B., Zhang, H.: Analysis of two parareal algorithms for time-periodic problems. SIAM J. Sci. Comput. **35**(5), A2393–A2415 (2013). DOI 10.1137/130909172

[7] Gander, M.J., Kulchytska-Ruchka, I., Niyonzima, I., Schöps, S.: A new parareal algorithm for problems with discontinuous sources. SIAM J. Sci. Comput. **41**(2), B375–B395 (2019). DOI 10.1137/18M1175653

[8] Gander, M.J., Kulchytska-Ruchka, I., Schöps, S.: A new parareal algorithm for time-periodic problems with discontinuous inputs. Domain Decomposition Methods in Science and Engineering XXV. Lecture Notes in Computational Science and Engineering. Springer (2019)

[9] Lions, J.L., Maday, Y., Turinici, G.: A parareal in time discretization of PDEs. Comptes Rendus de l'Académie des Sciences – Series I – Mathematics **332**(7), 661–668 (2001). DOI 10.1016/S0764-4442(00)01793-6

[10] Takahashi, Y., Tokumasu, T., Kameari, A., Kaimori, H., Fujita, M., Iwashita, T., Wakao, S.: Convergence acceleration of time-periodic electromagnetic field analysis by singularity decomposition-explicit error correction method. IEEE Trans. Magn. **46**(8), 2947–2950 (2010). DOI 10.1109/TMAG.2010.2043721

[11] Ulbrich, S.: Generalized SQP Methods with "Parareal" Time-Domain Decomposition for Time-Dependent PDE-Constrained Optimization, pp. 145–168. SIAM (2007). DOI 10.1137/1.9780898718935.ch7

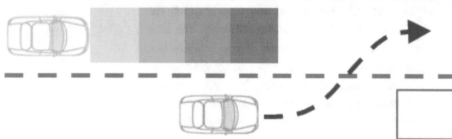

A cut-In scenario in autonomous driving

© Springer Nature Switzerland AG 2021
H. G. Bock et al. (eds.), *German Success Stories in Industrial Mathematics*,
Mathematics in Industry 35, https://doi.org/10.1007/978-3-030-81455-7_23

EFFICIENT VIRTUAL DESIGN AND TESTING OF AUTONOMOUS VEHICLES

A case study using multiobjective optimization

SEBASTIAN PEITZ
MICHAEL DELLNITZ
Institute for Industrial Mathematics,
Paderborn University

SEBASTIAN BANNENBERG
dSPACE GmbH, Paderborn

With the ever increasing capabilities of sensors and controllers, autonomous driving is quickly becoming a reality. This disruptive change in the automotive industry poses major challenges for manufacturers as well as suppliers as entirely new design and testing strategies have to be developed to remain competitive. Most importantly, the complexity of autonomously driving vehicles in a complex, uncertain, and safety-critical environment requires new testing procedures to cover the almost infinite range of potential scenarios. For this purpose, *scenario-based testing* is used as new test strategy where detailed physical models are used to simulate critical situations and test the vehicle behavior. As the computational capacities are quickly exceeded by naive testing, optimization techniques have to be used. In this context, multiobjective optimization allows us to identify critical testing situations while at the same time optimizing the system configuration.

The combination of multiobjective optimization algorithms developed by the Institute for Industrial Mathematics with dSPACE scenario-based testing solutions results in an intelligent and efficient approach for developing and testing functions for autonomous driving. The latest approach of the cooperation treats design and testing as a single process. This yields major improvements in the efficiency and speed of the development process of complex systems.

PARTNERS

SEBASTIAN BANNENBERG, dSPACE GmbH, Paderborn, Germany

Industrial challenge and motivation

The development of autonomous vehicles is a true game changer in the automotive and related industries, and experts agree that it will result in disruptive changes in the development process. One of the key challenges of autonomous vehicles – besides the design – is the necessity to develop and implement effective testing and validation processes, as safety is the central factor for a broad acceptance of this new technology.   One key part to be tested are the *electronic control units (ECUs)* of the vehicle, and because its functionalities are drastically increasing, countless driving situations have to be tested to ensure proper validation.   For this, the scale of $2\,000\,000$ test kilometers will very likely be required [10].   Even today, it is no longer practicable to reliably validate all functionalities based on prototypes and real vehicles alone [1, Chapter 3.4]. To this end, a large part of the tests has to be moved to the software side by means of simulation, which results in *virtual validation.*

In the past, testing was mostly performed *requirements-based*, i.e., the system under test had to be tested in a predefined catalog of requirements. In the context of autonomous driving, this is no longer sufficient due to the extreme complexity. Instead, it is necessary to complement the standard testing method with additional approaches such as *scenario-based testing* as part of the virtual validation. A scenario in the context of autonomous driving mainly consists of road information, driving maneuvers, and of environmental traffic.   The virtual vehicle under test (the *Ego-vehicle*) is placed into the scenario, and various *key performance indicators (KPIs)* are calculated from simulation data, evaluating for example the safety, the consumption, and the comfort.

The testing process has to be performed with respect to several higher-level criteria.   On the one hand, it must guarantee a reliable coverage of relevant traffic situations while taking into account the key performance indicators.   On the other hand, the computational effort has to be handled.   As the variation of parameters in the scenario results in an exponentially large number of tests, mathematical tools have to be developed to perform intelligent and

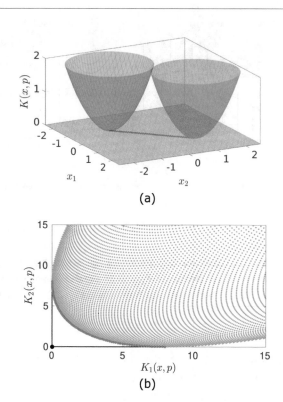

(a)

(b)

Figure 1: (a) The *Pareto set* (denoted in red) of an exemplary MOP with two internal parameters and two KPIs (here, the minimization of two quadratic functions). (b) The corresponding *Pareto front* (again in red) consists of the optimal compromises for which it is impossible to improve both objectives simultaneously.

efficient testing. This requires the use of algorithms from *multiobjective optimization* ([4], see also [3, 7, 6] for applications in the context of autonomous driving) and *sensitivity analysis*.   As a test tool supplier, dSPACE is developing an innovative solution for intelligent scenario-based testing that involves these algorithms (e.g., provided by institutes such as the IFIM).

Mathematical research

In a scenario-based design process, a parametrized driving scenario (with parameters $p \in \mathbb{R}^m$) is simulated using a detailed model of the vehicle dynamics as well as the environment. The vehicle behavior can be altered by varying certain internal

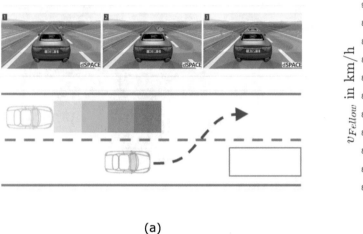

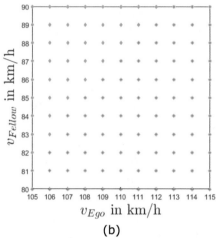

(a)   (b)

Figure 2: (a) Cut-In scenario, where the Ego vehicle (blue) on the left lane has to react to a sudden switching of lane of the Fellow vehicle (grey). (b) Evaluation of K_{safety} on a grid of parameter values. Every point in this figure represents one simulation. Crash cases ($K_{safety} = 0$) are marked red and non-crash cases ($K_{safety} > 0$) are marked green.

parameters (e.g., of the ECUs) – denoted by $x \in \mathbb{R}^n$ – and the results are evaluated via the *Key Performance Indicators (KPIs)* $K_1, \ldots, K_k$ with $K_i \colon \mathbb{R}^n \times \mathbb{R}^m \to \mathbb{R} \ \forall i = 1, \ldots, k$. When rephrasing this problem in terms of optimization, this results in a *parametric multiobjective optimization problem* with n free variables, m parameters, and k objectives:

$$\min_{x \in \mathbb{R}^n} K(x,p) = \min_{x \in \mathbb{R}^n} \begin{pmatrix} K_1(x,p) \\ \vdots \\ K_k(x,p) \end{pmatrix}. \quad \text{(MOP)}$$

A logical scenario with a fixed parameter p and specified key performance indicators is also called a *concrete test case* [9]. The solution to the problem (MOP) is the *Pareto set*, cf. [4] for a detailed introduction and an overview of numerical algorithms. This is the set of *optimal compromises*, i.e., of solutions where one objective can be improved only when accepting a trade-off in at least one other objective, see Figure 1. Knowledge of the entire Pareto set thus allows a developer to evaluate trade-offs between conflicting KPIs and to make a well-informed decision.

Example 1 (Testing an autonomous vehicle in the Cut-In scenario). In the *Cut-In scenario* (cf. Figure 2 (a) for an illustration), the *Ego-vehicle* is driving in the left lane of a two-lane highway – traveling at a higher speed than another vehicle (*Fellow-vehicle*) in the right lane. When the Ego is at almost the same height, the Fellow switches into the left lane, potentially causing an accident. The task of the autonomous vehicle is to detect these situations and to prevent collisions over a large range of parameters p by means of braking. In this setting, the parameters are the Ego-velocity v_{Ego} and the Fellow-velocity v_{Fellow} (with the ranges $v_{Ego} \in [105, 115]$km/h and $v_{Fellow} \in [80, 90]$km/h). The most relevant KPI in this scenario describes the safety of the vehicle and is defined by the minimum distance between the Ego and the Fellow ($d_{min}(v_{Ego}, v_{Fellow})$) captured during the simulation:

$$K_{safety} \colon \mathbb{R}^2 \to \mathbb{R},$$
$$K_{safety}(v_{Ego}, v_{Fellow}) = \max\{0, d_{min}(v_{Ego}, v_{Fellow})\}.$$

Other key performance indicators for this scenario can describe the comfort or the energy consumption. Based on K_{safety}, the *critical test cases* can be defined. These are the set of concrete test cases, separating the acceptable test cases ($K_{safety} > 0$) from the inacceptable ones ($K_{safety} = 0$). In Figure 2 (b), this is visualized with green and red dots.

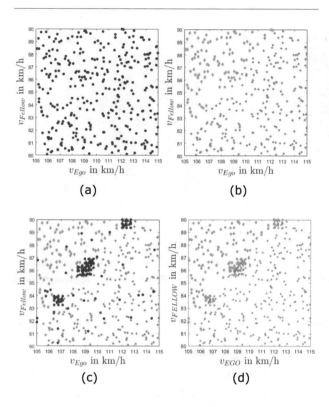

Figure 3: Iterative approximation of the entire set of critical test cases using a classical evolutionary algorithm. The blue points denote new simulations, and the green and red dots safe and unsafe test cases, respectively.

Identification of critical test cases

The first task is exclusively dedicated to identifying the critical test cases, i.e., the test cases separating the safe from the unsafe regions in the parameter domain. This can be implemented by solving single-objective optimization problems, using *evolutionary optimization* [8] to compute the entire set of critical test cases (cf. Figure 3 for an illustration) or a classical descend method such as *SQP* [5] for the *identification of the nearest critical test case (INCT)* of a given safe concrete test case $(\hat{x}, \hat{p})$ (with $K_{safety}(\hat{p}) > 0$). In the latter case, we keep the vehicle configuration $\hat{x}$ fixed and search for the closest unsafe test case p:

$$\min_{p} \|p - \hat{p}\|_2^2 \qquad \text{(INCT)}$$

$$\text{s.t.} \quad K_{safety}(p) = 0, \ K_{safety}(\hat{p}) > 0.$$

One exemplary solution of this problem is shown in Figure 4 (a), where $\hat{p}$ is marked by the black diamond and the optimum p^* by the red circle.

Robust multicriteria design

While the evolutionary approach yields a satisfactory approximation of the set of critical test cases – knowledge of which is crucial for autonomous driving – the second approach enables us to join the design and test process, thus yielding a very efficient tool for faster development cycles. To this end, we solve the parametric problem (MOP) for a moderate number of test cases $\hat{p}^1, \ldots, \hat{p}^N$, which yields the corresponding Pareto sets between the conflicting design criteria such as safety, energy efficiency, and driving comfort. In the next step, we compute the closest critical test case for all Pareto optimal configurations x^* in every test case $\hat{p}^i$ by solving problem (INCT). This allows us to define *safe regions* around every test case $\hat{p}^i$ within which a Pareto optimal design exists, cf. Figure 4 (b) and (c) for an illustration. This significantly helps in ensuring vehicle safety while preserving Pareto optimal performance. In particular, if we can cover the entire test space with safe regions, we can choose a Pareto optimal design parameter for every driving situation. This will allow us to adapt the vehicle configuration online, to ensure both safety and optimality.

Implementation

The collaboration between the Institute for Industrial Mathematics (IFIM) at Paderborn University and dSPACE began in 2017 with the question whether multiobjective optimization could support the scenario-based development and test process for highly complex systems, such as autonomous vehicles. The first results for efficient vehicle design using multiobjective optimization were obtained in a Master's thesis jointly supervised by dSPACE and the IFIM [2]. Central to this work was the combination of the problems (MOP) and (INCT) to join optimization and testing of autonomous vehicles. Further research was then concerned with increasing the numerical efficiency using surrogate

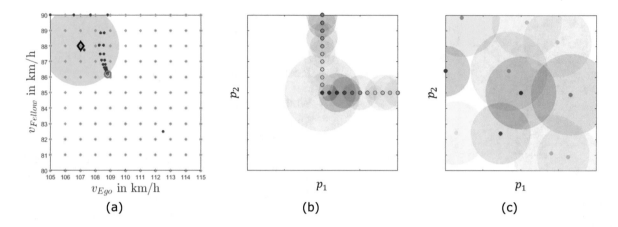

Figure 4: (a) Computation of the closest critical test case. The diamond marks the initial test case and the red circle marks the nearest critical concrete test case. (b, c) Illustration of the robust multicriteria design approach. The circles denote the safe regions around a parameter p within which a Pareto optimal design can be used.

modeling, for which again a student thesis was used as the starting point. Realizing the large potential of multiobjective optimization methods, the cooperation was significantly intensified at the beginning of 2019. At this point, the car manufacturer e.GO Mobile AG was involved as a customer and user of the software for the development and testing of autonomous electric vehicles. Since then, the cooperation has continued as part of the project *Simultanes Entwickeln und Testen von Cyber Physical Systems* (SET CPS, *Simultaneous Design and Testing of Cyber Physical Systems*) which is funded by EFRE.NRW (EFRE-0801346) with approximately one Million Euros.

Industrial relevance and summary

The customer basis of dSPACE includes automobile manufacturers from all over the world. dSPACE cooperates with these customers to significantly improve and accelerate the design and test process for autonomous driving. To this end, dSPACE offers an innovative and scalable tool chain for development and testing of complex systems that can be operated on a large scale in parallel in the cloud. As complexity is significantly increased for autonomous vehicles, smart and adaptive procedures are a key enabler to meet the great

challenge of autonomous driving. The selection of the necessary and sufficient scenarios is a key aspect in the scenario-based approach. Mathematical algorithms, e.g., from multiobjective optimization, are used for this purpose.

In the context of autonomous driving, the design and testing process have been handled mostly independently and in a consecutive manner. The collaboration between IFIM and dSPACE leads to creating tools that enable close integration of design and test, avoid unnecessary design loops, and significantly accelerate the entire development process. The first methods are currently prepared for the incorporation in the dSPACE solution for scenario-based testing, which results in a major improvement. In the future, the developed tools have the potential to significantly influence the continued development process of autonomous vehicles and to accelerate the time-to-market for new and innovative concepts.

Patents

» S. Bannenberg, R. Rasche, PCT Application PCT/EP2019/054149, 2019

References

[1] Aptiv, Audi, Baidu, BMW, Continental, Daimler, Fiat Chrysler Automobiles, HERE, Infineon, Intel, and

Volkswagen. *Safety first for automated driving*. Aptiv, Audi, Baidu, BMW, Continental, Daimler, Fiat Chrysler Automobiles, HERE, Infineon, Intel und Volkswagen, 2019.

[2] S. Bannenberg. Mengenorientierte mehrzieloptimierung und die identifikation kritischer testfälle im kontext des hochautomatisierten fahrens. *dSPACE GmbH*, 2017.

[3] M. Dellnitz, J. Eckstein, K. Flaßkamp, P. Friedel, C. Horenkamp, U. Köhler, S. Ober-Blöbaum, S. Peitz, and S. Tiemeyer. Multiobjective Optimal Control Methods for the Development of an Intelligent Cruise Control. In G. Russo, V. Capasso, G. Nicosia, and V. Romano, editors, *Progress in Industrial Mathematics at ECMI 2014*, pages 633–641, Taormina, Italy, 2017. Springer.

[4] M. Ehrgott. *Multicriteria optimization*. Springer Berlin Heidelberg New York, 2 edition, 2005.

[5] J. Nocedal and S. J. Wright. *Numerical Optimization*. Springer Series in Operations Research and Financial Engineering. Springer Science & Business Media, 2 edition, 2006.

[6] S. Ober-Blöbaum and S. Peitz. Explicit multiobjective model predictive control for nonlinear systems with symmetries. International Journal of Robust and Nonlinear Control, 2020 DOI: 10.1002/rnc.5281

[7] S. Peitz, K. Schäfer, S. Ober-Blöbaum, J. Eckstein, U. Köhler, and M. Dellnitz. A Multiobjective MPC Approach for Autonomously Driven Electric Vehicles. *IFAC PapersOnLine*, 50(1):8674–8679, 2017.

[8] D. Simon. *Evolutionary optimization algorithms*. John Wiley & Sons, 2013.

[9] M. Steimle, G. Bagschik, T. Menzel, J. Wendler, and M. Maurer. Ein beitrag zur terminologie für den szenarienbasierten testansatz automatisierter fahrfunktionen, 02 2018.

[10] W. Wachenfeld and H. Winner. The release of autonomous vehicles. In M. Maurer, J. C. Gerdes, B. Lenz, and H. Winner, editors, *Autonomous Driving: Technical, Legal and Social Aspects*, pages 425–449. Springer Berlin Heidelberg, 2016.

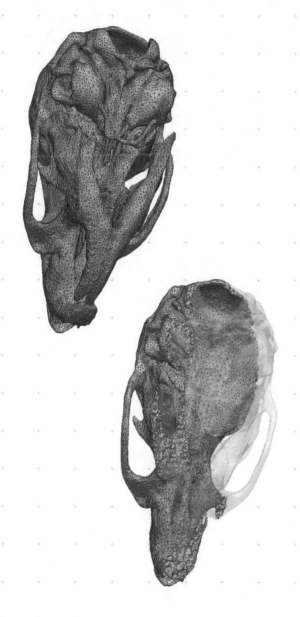

TetGen: Tetrahedral Mesh Generation
for Complex Simulations.

© Springer Nature Switzerland AG 2021
H. G. Bock et al. (eds.), *German Success Stories in Industrial Mathematics*,
Mathematics in Industry 35, https://doi.org/10.1007/978-3-030-81455-7_24

TETGEN: TETRAHEDRAL MESH GENERATION FOR COMPLEX SIMULATIONS

Efficient quality meshing for complex 3D simulations

Electromagnetic fields in magnetic resonance imaging, groundwater flows, charge carriers in semiconductors and numerous further spatially distributed physical processes have one thing in common: they can be modeled in a sophisticated and efficient way by means of partial differential equations. The solution for such complex mathematical equations can usually not be obtained as a mathematical formula. Therefore, computer-based approximation methods are applied. In order to calculate these, the respective spatial domain must be divided in an intelligent way into a finite number of simple subdomains (e.g. tetrahedra).

DR. JÜRGEN FUHRMANN
DR. HANG SI
WIAS Berlin

Industrial challenge and motivation

Hence, one of the pivotal core algorithms for computer simulations is mesh generation: the subdivision of spatial domains described by representations of their surfaces generated by CAD and other tools into a finite number of simplest elements. In the three-dimensional space, the simplest possible element for this purpose is a tetrahedron. These elements are used to form the basic structures for the computer representation of the fields of unknowns investigated during simulations. These fields can represent very different physical data, e.g. temperature distributions, flow velocities, particle concentrations or electrical charge.

Quality and efficiency of mesh generation directly affect the robustness and the resolution capabilities of computer simulations.

Figure 1:   TetGen: Tetrahedralization of the model of a cylinder head.

Mathematical research

Especially for three-dimensional geometries, the development of efficient mesh generation algorithms capable to deliver high quality meshes for complex domains as they occur in industrial applications is linked to many challenges leading to several mathematical and algorithmical problems. Here we list a few of them.

- For a given set of points, a tetrahedralization of this point set is a partition of the part of the thredimensional space occupied by this point set into tetrahedra such that each point corresponds to a vertex of this tetrahedralization. For a given point set, many different tetrahedralizations are possible. For any two given such tetrahedralizations, there may be a finite sequence of elementary local "moves" which allow to "morph" the first tetrahedralization into the second one. For a simple example, see Fig. 2. However it is not known if this speculation is true in general. This open question is an example for the lack of understanding of the fundamental combinatorial structures of triangulations of point sets [3].

- There exist comparably simple three-dimensional domains which cannot be subdivided into tetrahedra – tetrahedralized –

without introducing additional points like e.g. the Schönhard polytope, see Fig. 3. There are more of such rather elementary strutures, but there are no formal rules to recognize them. This problem is linked to fundamental questions of the geometry and topology of 3-manifolds which are not understood.

- Geometries to be tetrahedralized may have a number of important features which need to be preserved in the tetrahedralization process in order to allow for meaningul simulations, like e.g. the exact representation of certain points and edges in the discretization. For arbitrary 3D geometries there are no algorithms for meshing which guarantee the preservation of these features and whose functionality can be proven in a mathematically strict way.

- Also, the fields of unknowns to be obtained by a computer simulation may exhbit certain features like sharp fronts or limitations of values which can only represented accurately if the mesh is well adapted to them. Many of the resulting additional requirements on mesh generation are yet unsolved.

lem, bringing together results from such diverse fields of mathematics as computational geometry, differential geometry and knot theory provides a framework for creating new ideas and algorithmical approaches directed at the development of better algorithms. Following this philosophy, the mesh generation research project of the Weierstrass Institute led by Hang Si was able to deliver a number of profound insights and new algorithmical solutions related related to these questions:

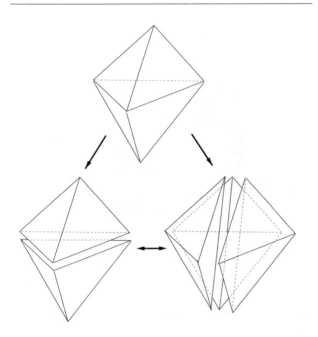

Figure 2: There are two ways to thetrahedralize five points setting up a triangular dipyramid. A "2-3-flip" is an elementary operation which "rewires" the tetrahedralization such that it contains either 2 or 3 tetrahedra, without affecting other tetrahedra in a possibly larger tetrahedralization.

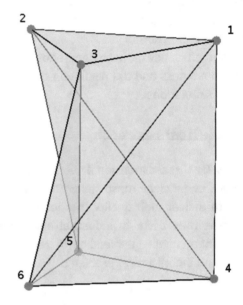

Figure 3: It is not possible to tetrahedralize the Schönhard polytope without introducing additional points.

- Another structural condition on thetrahedralizations of domains concerns the locations of the centers of the cirumspheres of the tetrahedra. For each tetrahedron, its circumsphere should contain the points of this one tetrahedron and no additional points from other tetrahedra. Also, the centers of all the circumspheres should be located within the domain or at its boundary. This so called boundary conforming Delaunay property enables certain discretization methods which allow to represent physical properties of the fields of unknowns in a particularly well defined manner. Combining this property with the requirement to use anisotropic tetrahedra in order to represent sharp fronts in an optimal way adds additional complexity.

At the Weierstrass institute, we believe that these open challenges can be successfully overcome only by the formalization of the unsolved questions and their formulation in precise mathematical terms. Given such a mathematically well formulated prob-

- It was possible to prove an algorithm to create boundary conforming Delaunay triangulations for domains described by polygons where the angle between two polygon facets is larger than $71°$. [10].

- New robust algorithms which have a high probability to create meshes which conform well to the input geometry have been derived in [9].

- In [11], new, fast algorithms for quality meshing have been derived.

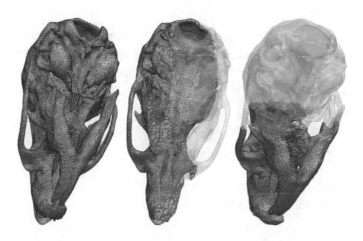

Figure 4:   TetGen: Tetrahedralization of the model of mouse skull.

Implementation

Accompanying the reserarch efforts described above, the algorithmic findings are implemented in the mesh generator TetGen. The strong mathematical backing reduces the role of heuristics and thus significantly increased robustness and efficiency of the code, see e.g. Figs 1, 4. This code holds an internationally leading position among three-dimensional mesh generation tools for academia and industry. Available under the open source Affero Gnu Public License v3.0 (AGPL) [1] , TetGen [8] finds widespread use in academic research:

- TetGen can be used to create discretization grids by the open source finite element simulation environment FEniCS [2].

- TetGen was used by astrophysicists to support finite element analysis of the stability of a comet [4].

- Geophycisists use TetGen to mesh complicated geological structures for the solution of electromagnetic field problems [6].

- TetGen is used in biomedical research to insert bone samples into anatomic models of the human body [5]

- In the field of biomolecular research, TetGen is used to create three-dimensional models of ion channels [7].

A particular advantage of this open source based distribution strategy is the huge amount of feedback which helps to reveal implementation bugs and provides new challenging problems to solve. The copyleft associated with the AGPL license ensures that the research simulation tools using TetGen are distributed to the community under similar conditions.

Industrial relevance and summary

To companies that like to avoid the copyleft commitment resulting from the AGPL in order to allow for a commercial use, WIAS provides commercial licenses for a fee. The revenues contribute to the resources available to the project and together with citation metric provide an addional indicator on its standing. TetGen is used by companies all over the world. Within the last ten years, 26 such commercial licenses have been sold, e.g. to

- JCMwave for photonics simulations,

- Wolfram Research for Mathematica,

- Google for image processing,

- Side Effects for Virtual reality and gaming engines,

- Tera for calculation of electromagnetic fields and temperature distributions in complex electric components

References

[1] GNU Affero General Public License. URL: https://www.gnu.org/licenses/agpl-3.0.en.html.

[2] M. Alnæs, J. Blechta, J. Hake, A. Johansson, B. Kehlet, A. Logg, C. Richardson, J. Ring, M. E. Rognes, and G. N. Wells. The fenics project version 1.5. *Archive of Numerical Software*, 3(100), 2015.

[3] J. A. De Loera, J. Rambau, and F. Santos. *Triangulations Structures for algorithms and applications*. Springer, 2010.

[4] M. Hirabayashi, D. J. Scheeres, S. R. Chesley, S. Marchi, J. W. McMahon, J. Steckloff, S. Mottola, S. P. Naidu, and T. Bowling. Fission and reconfiguration of bilobate comets as revealed by 67p/churyumov–gerasimenko. *Nature*, 534(7607):352–355, 2016.

[5] P. Kadleček, A.-E. Ichim, T. Liu, J. Křivánek, and L. Kavan. Reconstructing personalized anatomical models for physics-based body animation. *ACM Transactions on Graphics (TOG)*, 35(6):1–13, 2016.

[6] J. Li, C. G. Farquharson, and X. Hu. 3d vector finite-element electromagnetic forward modeling for large loop sources using a total-field algorithm and unstructured tetrahedral grids. *Geophysics*, 82(1):E1–E16, 2017.

[7] X. Liu and B. Lu. Incorporating born solvation energy into the three-dimensional poisson-nernst-planck model to study ion selectivity in kcsa k+ channels. *Physical Review E*, 96(6):062416, 2017.

[8] H. Si. TetGen – a quality tetrahedral mesh generator and a 3D Delaunay triangulator. URL: https://tetgen.org.

[9] H. Si. TetGen, a Delaunay-based quality tetrahedral mesh generator. *ACM Transactions on Mathematical Software (TOMS)*, 41(2):1–36, 2015.

[10] H. Si, K. Gärtner, and J. Fuhrmann. Boundary conforming Delaunay mesh generation. *Computational Mathematics and Mathematical Physics*, 50(1):38–53, 2010.

[11] H. Si and J. R. Shewchuk. Incrementally constructing and updating constrained Delaunay tetrahedralizations with finite-precision coordinates. *Engineering with Computers*, 30(2):253–269, 2014.

In this exhibit the solver WORHP computes optimal trajectories based on the modeled landscape.

WORHP: DEVELOPMENT AND APPLICATIONS OF THE ESA NLP SOLVER

An NLP solver for the European Space Agency ESA

WORHP is a mathematical software library for solving continuous large-scale nonlinear optimization problems numerically [Bueskens2013]. The acronym WORHP stands for *We Optimize Really Huge Problems*, its primary intended application.

What does „huge" mean? Today, WORHP can solve problems with more than 1 billion degrees of freedom and 2 billion constraints.

Rumor has it that a first draft of our optimization software was created on a beer mat. Based on that, in a secluded environment, a first version was developed in a coding bootcamp by a team of 17 people.

In the further development period sprints were planned again and again, where the code was intensively developed using pair programming.

Over the years WORHP turned into a state-of-the-art solver for nonlinear optimization problems (NLP) and was declared as the official ESA NLP solver.

It has a growing community of users from over 35 countries and is used succesfully in several academic and industrial projects.

Optimization problems occur in many applications through all branches of industry. It turned out that projects from these sectors are highly suited for applying WORHP:

+ Especially in **aerospace**, where relatively accurate models exist, the desired spatial or temporal resolution leads to large-scale NLPs.

+ For **autonomous systems**, e.g. assisted driving, fast optimization cycles are crucial whether the system can be used in real-time applications.

+ In the **energy sector** huge sets of measurement data are available to identify model parameters. This calls for efficient algorithms which take into account the computational costs of any function evaluation.

MATTHIAS KNAUER
CHRISTOF BÜSKENS
Center for Industrial Mathematics, University of Bremen

PARTNERS

BMWi (German Federal Ministry of Economics and Technology) | TEC-EC Control Division of ESA (European Space Agency)

Industrial challenge and motivation

Since many years modelling and simulation are proven techniques in understanding industrial processes. Models, either physically justified or purely data based, could be generated and iteratively increased in their precision, as huge amounts of data are available from the industrial application. If differences between a monitored process in the real world and its corresponding simulation occur, the source for these failures can be tracked down early to avoid larger hazards.

However, these well established modelling concepts and simulation techniques are used rarely to improve or even optimize the process itself. As models grew more and more complex, the pure computation times for single simulations increased even though processing power was improving continuously. This keeps users from connecting their model with an optimization framework.

Both static and dynamic models of industrial processes can be written as systems of equations (and of course inequalities). During the numerical simulation the lengthy task of solving these systems has to be performed. If an optimization consists of invoking several simulations, this nested approach is indeed time consuming.

On the other hand, the system could be solved while simultaneously adjusting selected system parameters to optimize for a given goal. This typically requires that also the outcome of a simulation has to be considered as variable. Hence, all system parameters and simulation outcomes are used as optimization variables, leading to huge problem dimensions.

The working group Optimization and Optimal Control at the Center of Industrial Mathematics at the Universität Bremen focuses on the development and application of mathematical software to solve these large-scaled optimization problems.

Mathematical research

Though WORHP can solve problems emerging from the most different applications, all can be led back to the same mathematical formulation of nonlinear programming problems:

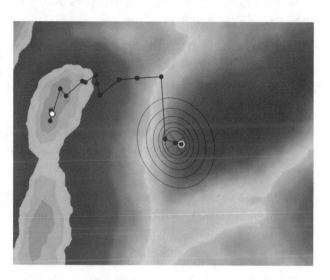

Figure 1: The task of finding the lowest point in a landscape (as shown in this topopgraphic map) can be used to illustrate the SQP method.

$$\min_{x \in R^n} \quad f(x)$$
$$\text{s.t.} \quad g(x) \leq 0$$
$$h(x) = 0$$

Herein, a (possibly very large) set of variables x has to be found, such that an objective function $f(x)$ is minimized while fulfilling (possibly very large sets of) inequality and equality constraints, defined by the sets of functions $g(x)$ and $h(x)$. Generally, each function only depends on a small number of variables. Whenever derivatives of the functions are needed, the sparsity patterns of the resulting derivative matrices can be exploited.

WORHP employs a sequential quadratic programming method (SQP), to iteratively approximate the NLP by a quadratic problem (QP). In Fig. 1 the QP for the first iteration is visualized as height lines. These height lines form a paraboloid, from which the minimum can easily be computed. In this case the found point is fully accepted and the NLP will be approximated around this point by a new QP, until eventually the (local) minimum of the NLP is found.

Even though SQP methods are well documented in

the literature, a competitive implementation requires a lot of widely spread research and endurance. The scientific contributions of our PhD students helped us to make WORHP a unique solver:

- In industrial projects analytic derivatives often can not be provided. However, approaches from group theory can be used for an efficient calculation of numerical derivatives [2].

- Results from parametric sensitivity analysis can be used during the optimization process to improve convergence [7].

- The improvements of the architecture and the implementation of WORHP itself was the subject in [9].

- Sparse BFGS methods to approximate second derivatives, which use only limited memory, are discussed in [8].

- A range of concepts to improve convergence are examined in [1].

- An interior-point method as an alternative to the describe SQP method was conceived in [5].

Implementation

The development of WORHP was initiated by the European Space Agency ESA in 2006. When they claimed that there is a need for a European solver for optimization problems, their main interest was to achieve independence from non-European software. The engineers at ESA formulated the goal for the development of the new solver to especially address optimization problems from real-life applications. They had deep space maneuvers with electric propulsion in mind, where some hundred thousand of optimization variables and constraints have to be considered. Further, the new solver should be competitive regarding the robustness of the solution found, and robustness was considered more important than computing speed. To simplify the usage of the solver, its output, especially the error messages, should be comprehensible also to non-mathematicians.

The software was never intended as an experimental research project, but for usage in everyday industrial life – meeting the strict industry standards from the European Cooperation for Space Standardization. To reach this high goal, the complete structure and the methods of WORHP were designed on the drawing board first and an extensive literature and method research was done.

In order to develop a solver which will be used by a wide group of users, all interfaces and concepts for the new solver were discussed and decided together with our partners at ESA. During the first years, the development was additionally accompanied by an industrial partner to implement independent test runs.

One important decision was the programming language to be used. In our benchmark test Fortran outperformed all other languages when it came to pure number crushing. Hence the core of WORHP is developed in Fortran and various interfaces to C, C++, Python, AMPL, and MATLAB ensure a wide range of applicability.

During the first project phase, which was funded by the German Aerospace Center DLR, and later continued by ESA fundings, we worked together with Prof. Matthias Gerdts from the University Hamburg. His group developed a QP solver based on an interior-point method, while the group of Prof. Christof Büskens implemented an SQP method. Until today WORHP was improved various times. An alternative QP solver was added and besides the SQP method there also exists a pure interior-point method.

Since the solution of large optimization problems requires the handling of large matrices, an efficient implementation has to exploit sparse matrix structures. The necessary linear algebra routines are already available in standard libraries like SuperLU and MUMPS. Back then we decided together with ESA to use these and not to implement our own. Since some years, a special version of MA97 is included in WORHP to perform linear algebra operations in a highly parallelized way.

Over the years, the development of WORHP was funded by several projects. In the side project Sentinel a "watchdog" should be implemented to analyse the situation in case WORHP would not terminate successfully. The intention was to give helpful hints to the user. However, these hints worked out so well, and could be automated so easily,

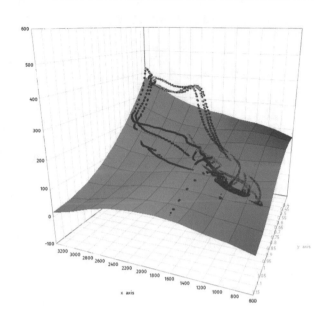

Figure 2: Automated parametrization of characteristic maps is used in automotive companies.

such that they are now integated in WORHP and work independently from the user's experience.

A real strength of WORHP is its flexibility as the solver can be tweaked with over 200 parameters. This means, all parameters, which define the behaviour of any numerical method, are available for manipulation by the more experienced users in order to build their own problem-dependent solver. Finding the best set of parameters for a class of optimization problems is an optimzation problem on its own. The open access to the parameters helps to automate this process.

All users are encouraged to work with the reverse communication interface, where all parameter settings and even the problem itself can be changed during runtime. Overall, WORHP features a modular structure to be able to flexibly respond to change requests during our development cycles. This so-called unified solver interface enables a massively flexible communication interface.

During the development we used the AMPL CUTEr test set as a benchmark test for the robustness of the solver. We are happy to know that WORHP is able to solve 100 percent of all solvable problems.

Industrial relevance and summary

As already described, the routines of WORHP are encapsuled, but with a gigantic open interface. At ESA's request, WORHP is not available as open source software, but rather as software for which maintenance work can be guaranteed for another 10 years.

In the initial planning, WORHP should actually only be available as a commercial product. We decided, however, that the software may and should be used permanently for research purposes free of charge (as a binary file).

Since the release of the first version of WORHP in 2011 additional tools have been developed:

- **WORHP Zen** is a module that enables the computation of a parametric sensitivity analysis. Now, not only the optimal solution is computed, but also additional information is generated, on how the optimal solution would change, if system parameters are varied [6].

- **TransWORHP** the companion transcription method to WORHP and was especially developed for optimal control problems.   Here, an (infinite dimensional) optimal control problem is discretized to a large, but finite dimensional NLP [4].

- **WORHP Lab** is a graphical user interface with a reduced functionality, which is especially suited for education purposes. We also use it to explain our ideas of optimization to new industrial contacts in hands-on workshops.

For our group, WORHP serves as a door opener for several industrial contacts. In the last years we learned that WORHP is especially powerful in these contexts:

- WORHP can be used in techniques like model predictive control (MPC) to generate solutions **particularly fast** [3].

- Optimization with PDEs generates extremely **high dimensional**, but well structured problems.

- In industrial applications the **robustness** ensures that similar problems are solved reliably again and again.

The success of WORHP is not that one problem has been solved, but in the plethora of solvable problems.

References

[1] S. Geffken. *Effizienzsteigerung numerischer Verfahren der nichtlinearen Optimierung*. PhD thesis, Universität Bremen, 2017.

[2] P. Kalmbach. *Effiziente Ableitungsbestimmung bei hochdimensionaler nichtlinearer Optimierung*. PhD thesis, Universität Bremen, 2011.

[3] M. Knauer and C. Büskens. Processing User Input in Tracking Problems using Model Predictive Control. *IFAC-PapersOnLine*, 50(1):9846–9851, 2017.

[4] M. Knauer and C. Büskens. *Modeling and Optimization in Space Engineering*, volume 144 of *Springer Optimization and Its Applications*, chapter Real-Time Optimal Control Using TransWORHP and WORHP Zen, pages 211–232. Springer, 2019.

[5] R. Kuhlmann. *A Primal-Dual Augmented Lagrangian Penalty-Interior-Point Algorithm for Nonlinear Programming*. PhD thesis, Universität Bremen, 2018.

[6] R. Kuhlmann, S. Geffken, and C. Büskens. WORHP Zen: Parametric Sensitivity Analysis for the Nonlinear Programming Solver WORHP. In N. Kliewer, J. F. Ehmke, and R. Borndörfer, editors, *Operations Research Proceedings 2017*, pages 649–654. Springer, 2018.

[7] T. Nikolayzik. *Korrekturverfahren zur numerischen Lösung nichtlinearer Optimierungsprobleme mittels Methoden der parametrischen Sensitivitätsanalyse*. PhD thesis, Universität Bremen, 2012.

[8] S. Rauski. *Limited Memory BFGS method for Sparse and Large-Scale Nonlinear Optimization*. PhD thesis, Universität Bremen, 2014.

[9] D. Wassel. *Exploring novel designs of NLP solvers: Architecture and Implementation of WORHP*. PhD thesis, Universität Bremen, 2013.

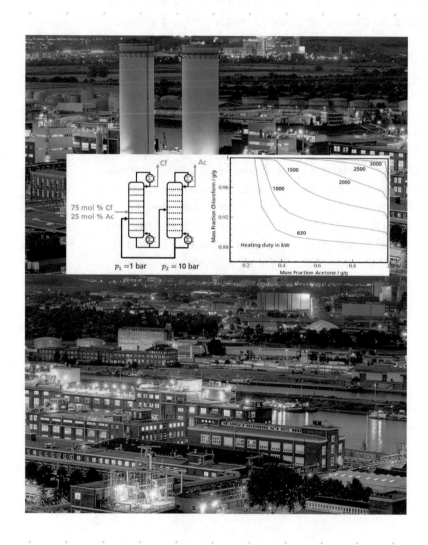

© Springer Nature Switzerland AG 2021
H. G. Bock et al. (eds.), *German Success Stories in Industrial Mathematics*,
Mathematics in Industry 35, https://doi.org/10.1007/978-3-030-81455-7_26

NON-CONVEX PARETO SET NAVIGATION

Decision support in chemical industry and medicine

Many real world applications require the solution of multi-objective optimization problems. Exploration of the corresponding set of best compromises, i.e., the Pareto set, supports the decision making process [Allmendinger2017, Hwang1979]. Determination and exploration of Pareto sets for non-convex objectives are in particular challenging and worthy of research. Fraunhofer ITWM develops software solutions that connect non-convex Pareto set exploration with new mathematical methods, two of which [Nowak2019, ScheE-tAl-JoGO2015], in particular, have made an impact on the real world and are presented in this article.

The first method is generic and relies on an approximation of the Pareto front, see Figure 2. This method is especially suited for black box models and models that are computationally challenging but allow the determination of Pareto points. It was first applied in chemical industry and continues to find further applications. A non-convex industrial example follows this methods description. The second method addresses an application with specific practical demands on usability and visualization. It was developed for interactive multi-criteria planning in clinical radiotherapy and is illustrated by a real case study.

DIMITRI NOWAK
ALEXANDER SCHERRER
MICHAEL BORTZ
KARL-HEINZ KÜFER
**Fraunhofer Institute for
Industrial Mathematics (ITWM)**

PARTNERS

NORBERT ASPRION, **BASF SE, 67056 Ludwigshafen**

Industrial challenge and motivation

At Fraunhofer ITWM, decision support for multi-objective optimization problems can be described as a three stage process. If the problem is computationally not too challenging, the first two steps can be skipped. They construct an approximation of the Pareto set to be explored instead. The actual exploration happens in the third stage.

The first stage reduces the Pareto set to a finite number of representatives with good coverage by means of in-house algorithms such as Sandwiching [18], Hyperboxing [19] or a combination of both [4]. The second stage combines the reduced set for constructing an approximation of the Pareto front. For convex problems, this can be achieved by making use of the convex hull of the representatives [13, 14]. In case of non-convexity, the approximation is constructed based on the Delaunay triangulation of aligned Pareto front representatives [15]. This idea integrates the use of neighbouring Pareto front information and provides good results in competitive time for up to 10 objectives. Similar ideas with stronger conditions considerably slow down as early as 4 objectives have been reached [8]. The final

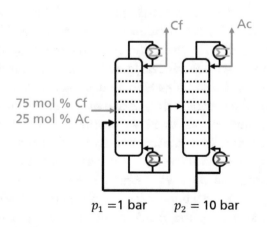

75 mol % Cf
25 mol % Ac

Cf   Ac

$p_1 = 1$ bar $p_2 = 10$ bar

Figure 1: Flowsheet for the thermal separation of Chloroform (Cf) and Acetone (Ac) by a pressure swing distillation.

stage is navigation [13, 14], which allows the

decision maker to perform two actions to explore the Pareto set or the approximation. One action is selection. The decision maker can change either one of the objectives or one of the parameters to certain value. The other action is restriction. The decision maker can bound the feasible area of either an objective or parameter to the range that he or she desires to explore further. For white box models, both actions result in nonlinear optimization runs. For the approximation of a non-convex Pareto set, the navigation has to be performed on a triangulation that approximates the Pareto front. For good performance, it makes use of ray tracing techniques and kd-tree data structures [15]. Ray tracing keeps the decision maker on the triangulation during the selection. Restriction makes use of bounding boxes and kd-tree structures to quickly access the feasible area in terms of linear optimization runs.

Ray tracing technique

The idea of ray tracing is based on the perspective projection and, thus, it is very (almost half a millennium) old [9]. In computer science, ray tracing is typically used in combination with kd-tree structures to quickly traverse numerous objects [7]. For non-convex Pareto set navigation [15], the selection action applies the perspective projection in direction of uniform improvement to stay on the Pareto front triangulation. It is the use of kd-tree structures that reduces the complexity of selection to $O(\log(N))$, where N is the number of triangles. Sometimes, though, ray tracing misses the triangulation or the triangle hit is infeasible due to some preceding restriction action. Then, linear optimization runs have to be performed solving:

$$d \quad \to \quad \min_{d,y} \quad \text{s.t.} \tag{1}$$
$$d \quad \geq \quad |y_i - p_i| \text{ for all } i \neq k$$
$$y_k \quad = \quad p_k$$
$$y \quad \in \quad \Delta$$

The result of (1) is the uniformly closest combination of Pareto points on the triangle Δ to the selected objective values p, given that the decision maker has changed objective k to value p_k, see Figure 2. This combination yields a corresponding solution

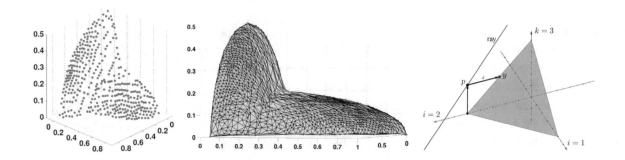

Figure 2: Three stages of a generic non-convex decision support example (from left to right): The first stage applies Hyperboxing for good coverage of the Pareto front. The second stage constructs an approximation in terms of Delaunay triangulation. The third stage uses ray tracing for navigation.

inside the Pareto set approximation. For parameter selection and restriction, optimization problem (1) can be extended resulting in linear optimization runs, as well.

Good performance of ray tracing depends on fast triangulation in the second stage of decision support. For the computation of a Delaunay triangulation, Quickhull [2] computes the convex hull of K projected Pareto points of dimension n with worst case complexity $O(K \log K)$ for $n \leq 3$ and $O\left(K^{\lfloor \frac{n}{2} \rfloor}\right)$ for $n \geq 4$. This is very competitive and promises good performance for up to $n = 10$ objectives.

Decision support in chemical engineering

In a series of bilateral collaboration projects, the worldwide largest chemical company, BASF, and Fraunhofer ITWM implemented a decision support framework to improve the design and operation of chemical production plants. Due to the strong nonlinearities in the underlying model equations, one generally has to expect non-convex Pareto sets, which the engineer can explore using non-convex navigation techniques. A particularly simple example for the occurrence of non-convex Pareto sets is sketched in this section. For further details, the interested reader is referred to [4].

Figure 1 shows a flowsheet consisting of two distillation columns designed for a thermal separation of a binary mixture consisting of 75 mol % Chloroform and 25 mol% Acetone. These two substances form a heavy boiling azeotrope at 65 mol % Chloroform at 1 bar. This means that in the first distillation column, a top stream which is rich in Chloroform could be achieved, while the limit for the Acetone concentration in the bottom product is 35 mol%. The higher pressure in the second column shifts the azeotrope towards higher Chloroform concentrations, so that an Acetone-rich product stream can be withdrawn at the top. The bottom is recycled. This process is simulated using an equilibrium-stage model with the BASF inhouse flowsheet simulator CHEMASIM. Of particular interest for the process engineer is the tradeoff between the two product concentrations and the heating duty in the columns. The concentrations of the two substances in the respective product streams should be as high as possible, while at the same time keeping the heating duty, i.e. the energy consumption, as small as possible. Four design parameters were chosen, namely the split and reflux ratios in the two columns. The result of the Pareto-optimization is shown is Figure 3. Here, Pareto sets for different heating duties are presented, obtained by maximizing the mass fractions of the products in their respective streams. It turns out that below a certain energy threshold, the Pareto sets show non-convex behaviour. By using the techniques sketched above , these can be navigated by the user in real time.

Radiotherapy planning

Radiotherapy (RT) is the most important option in oncology besides surgery and chemotherapy. Radiation emitted by external beams creates a dose distribution $\mathbf{d} = (d_k)_k$ represented on some volume grid of the body tissue. Modulation of beam intensities $\mathbf{x}$ (IM) allows for conformation of high doses to the tumor structures and wide sparing of nearby healthy tissue. The several planning goals of covering the tumor structures with sufficiently high doses and sparing the healthy structures with ideally low doses turns IMRT planning into a multi-criteria optimization problem [6]. Modern IMRT planning features real-time exploration of a convex hull of representative plans [11], which also makes extensive use of the dose volume histogram (DVH). The DVH summarizes the dose distribution on each planning structure with a curve, whose points indicate the shares of the structure receiving at least a certain dose value $\bar{d}$. A DVH criterion for a planning structure specified by its volume shares v_k of the grid cells thus reads

$$f_{\bar{d}}(\mathbf{d}(\mathbf{x})) = \sum_k v_k \cdot 1_{[\bar{d},\infty)}(d_k(\mathbf{x})) \qquad (2)$$

This criterion is isotone and connected and facilitates invex approximation [5][16]. Exploration of the convex plan set comprises a lot of local modifications on the dose-volume curves as and when needed. The characteristics of multi-criteria DVH-based IMRT planning are thus the absence of DVH criteria during Pareto approximation, their repeated modification during Pareto exploration and the need for real-time exploration with local search methods. Figure 4 (left) shows an IMRT planning case of head-neck cancer with the tumor structures GTV and PTV surrounded by left and right parotid gland and spinal cord. Therapy success requires wide dose coverage of GTV and PTV with doses around 60Gy. The current plan already comes close to these requirements, see Figure 4 (middle), but features a poor sparing of the parotid glands. In order to improve on the right gland, the physician specifies a DVH criterion of type (2) at $\bar{d} = 43$Gy by clicking on the corresponding curve point. This click triggers a nonlinear optimization run for the criterion value conducted on the convex hull of plans

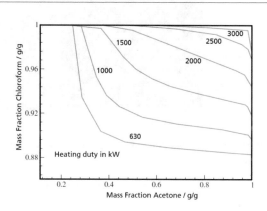

Figure 3: Pareto sets for different heating duties, obtained by maximizing the mass fractions of the products in their respective streams.

with Wolfe's method of reduced gradients [16][3]. This fast optimization run instantaneously yields a quasi-continuous trajectory of interpolated plans in the convex hull of representative plans, see Figure 4 (right), and a corresponding available value range displayed as colored interval on the vertical criterion axis, see Figure 4 (middle). The physician now modifies the current plan by moving the control element on this axis inside the range towards the desired value, while the DVH undergoes permanent updates and thereby provides immediate feedback on this selection operation. In order to preserve the achieved improvements before proceeding with the next DVH criterion, the physician may then restrict the marker position from above. DVH-based exploration of the convex hull of IMRT plans comprises a number of such selections and restrictions, possibly in combination with some extension of the available plan set by means of goal programming [12]. This functional and visual concept is covered by several filed and pending national patents and part of a commercial IMRT planning system [17][20].

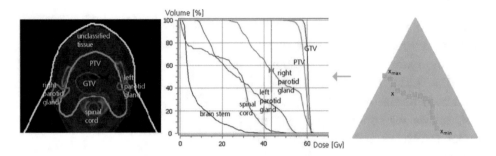

Figure 4: Left: Transversal view on body tissue with planning structures; Middle: DVH with criterion; Right: convex hull with trajectory of plans

References

[1] R Allmendinger, M Ehrgott, X Gandibleux, MJ Geiger, K Klamroth, M Luque: *Navigation in multiobjective optimization methods*. Journal of Multi-Criteria Decision Analysis 24(1-2):57-70 (2017)

[2] CB Barber, DP Dobkin, H Huhdanpaa: *The Quickhull Algorithm for Convex Hulls*. ACM Transactions on Mathematical Software 22(4):469-483 (1996)

[3] M Bazaraa, HD Sherali, CM Shetty: *Nonlinear Programming—Theory and Algorithms*. Wiley (1993)

[4] M Bortz, J Burger, N Asprion, S Blagov, R Böttcher, U Nowak, A Scheithauer, R Welke, KH Küfer, H Hasse: *Multi-criteria optimization in chemical process design and decision support by navigation on Pareto sets*. Computers & Chemical Engineering 60:354-363 (2014)

[5] BD Craven, BM Glover: *Invex functions and duality*. Journal of the Australian Mathematical Society A39:1–20 (1985)

[6] HW Hamacher, KH Küfer: *Inverse radiation therapy planning - a multiple objective optimization approach*. Discrete Applied Mathematics 118:145-161 (2002)

[7] M Hapala M, V Havran: *Review: Kd-tree Traversal Algorithms for Ray Tracing*. Computer Graphics Forum 30(1):199-213 (2011)

[8] M Hartikainen, K Miettinen, MM Wiecek: *Constructing a Pareto front approximation for decision making*. Mathematical Methods of Operations Research 73(2):209–234 (2011)

[9] GR Hofmann: *Who invented ray tracing?*. The Visual Computer 6(3):120-124 (1990)

[10] CL Hwang, AS Masud: *Multiple Objective Decision Making — Methods and Applications: A State-of-the-Art Survey*. Mathematical Methods of Operations Research. Springer, Berlin (1979)

[11] H Küfer, M Monz, A Scherrer, P Süss, F Alonso, AS Azizi Sultan, T Bortfeld, C Thieke: *Multicriteria optimization in intensity modulated radiotherapy planning*. PM Pardalos, HE Romeijn (ed.): *Handbook of Optimization in Medicine* (5):123-168, Kluwer (2009)

[12] K Miettinen: *Nonlinear Multiobjective Optimization*. Kluwer, Dordrecht (1999)

[13] M Monz: *Pareto Navigation – interactive multiobjective optimisation and its application in radiotherapy planning*. PhD thesis, Technical University Kaiserslautern (2006)

[14] M Monz, KH Küfer, TR Bortfeld, C Thieke: *Pareto navigation-algorithmic foundation of interactive multi-criteria IMRT planning"*. Physics in Medicine and Biology 53(4):985 (2008)

[15] D Nowak, KH Küfer: *A Ray Tracing Technique for the Navigation on a Non-convex Pareto Front. arXiv preprint arXiv:2001.03634*, 2019.

[16] A Scherrer, F Yaneva, T Grebe, KH Küfer: *A new algorithmic approach for handling DVH criteria in interactive IMRT planning*. Journal of Global Optimization 61(3):407-428 (2015)

[17] A Scherrer, KH Kuefer, P Suess, M Bortz: *Navigable presentation of a variety of solutions for therapy plans*. Patent Application WO/2013/093852 A1 (2015)

[18] JI Serna: *Approximating the Nondominated Set of R+ convex Bodies*. Master thesis, Technische Universität Kaiserslautern (2008)

[19] K Teichert: *A hyperboxing Pareto approximation method applied to radiofrequency ablation treatment planning*. PhD thesis, Technical University Kaiserslautern (2013)

[20] Varian Medical Systems: *Eclipse Feature Sheet*. Whitepaper Report (2018)

Correction to: Capacity Evaluation for Large-Scale Gas Networks

Martin Schmidt, Benjamin Hiller, Thorsten Koch, Marc E. Pfetsch,
Björn Geißler, René Henrion, Imke Joormann, Alexander Martin,
Antonio Morsi, Werner Römisch, Lars Schewe, Rüdiger Schultz,
and Marc C. Steinbach

Correction to:
Chapter "Capacity Evaluation for Large-Scale Gas Networks" in: H. G. Bock et al. (eds.),
German Success Stories in Industrial Mathematics, **Mathematics in Industry 35,**
https://doi.org/10.1007/978-3-030-81455-7_27

1. The original version of the book was inadvertently published without the following chapter co-author names and affiliation details in webpage of chapter 5, which have now been updated.
   - Björn Geißler - Adams Consult GmbH & Co. KG, Büttelborn
   - René Henrion - Weierstrass Institute for Applied Analysis and Stochastics, Berlin
   - Imke Joormann - TU Braunschweig, Institute for Mathematical Optimization, Braunschweig
   - Alexander Martin - (a) Friedrich-Alexander-Universität Erlangen-Nürnberg, Erlangen-Nürnberg (b) Fraunhofer Institute for Integrated Circuits IIS, Erlangen
   - Antonio Morsi - Adams Consult GmbH & Co. KG, Büttelborn
   - Werner Römisch - Humboldt-University Berlin, Institute of Mathematics, Berlin
   - Lars Schewe - University of Edinburgh, School of Mathematics, Edinburgh
   - Rüdiger Schultz - University of Duisburg-Essen, Faculty of Mathematics, Duisburg-Essen
   - Marc C. Steinbach - Leibniz Universität Hannover, Institute of Applied Mathematics, Hannover.

Furthermore, the affiliations of the co-authors "Martin Schmidt, Benjamin Hiller, Thorsten Koch, Marc E. Pfetsch" were incorrect, which have been now corrected.
   Correct Affiliations are:
   - Martin Schmidt - Trier University, Department of Mathematics, Trier
   - Benjamin Hiller - atesio GmbH, Berlin
   - Thorsten Koch - (a) TU Berlin, Chair for Software and Algorithms for Discrete Optimization, Berlin (b) Zuse Institute Berlin, Berlin
   - Marc E. Pfetsch - Technische Universität Darmstadt, AG Optimierung, Darmstadt,

---

The updated version of this chapter can be found at
https://doi.org/10.1007/978-3-030-81455-7_5

Printed in the United States
by Baker & Taylor Publisher Services